This textbook presents a broad coverage of atmospheric physics. It assumes a basic knowledge of undergraduate-level physics and mathematics but begins at a lower level than most other textbooks and shows how physics helps us to understand many important aspects of atmospheric behaviour.

An Introduction to Atmospheric Physics presents a thorough treatment of atmospheric thermodynamics, radiative transfer, atmospheric fluid dynamics and elementary atmospheric chemistry. Armed with an understanding of these topics, the interested student will be able to grasp the essential physics behind issues of current concern, such as the amplification of the greenhouse effect and associated questions of climatic change, the Antarctic ozone hole and global depletion of ozone, as well as more familiar processes such as the formation of raindrops and the development of weather systems.

This book is an excellent introductory text for intermediate-to-advanced undergraduates studying atmospheric physics as part of physics, meteorology and environmental science courses. It will also be useful for graduate students studying atmospheric physics for the first time and for students of applied mathematics, physical chemistry and engineering who have an interest in the atmosphere.

David G. Andrews is Reader in Physics at Oxford University. He is a Fellow of the Royal Meteorological Society and a Member of the Institute of Physics and American Meteorological Society. His previous publications include *Middle Atmosphere Dynamics* (with J. R. Holton and C. B. Leovy, Academic Press, 1987).

An Introduction to Atmospheric Physics

David G. Andrews

PUBLISHED BY THE PRESS SYNDICATE OF THE UNIVERSITY OF CAMBRIDGE
The Pitt Building, Trumpington Street, Cambridge, United Kingdom

CAMBRIDGE UNIVERSITY PRESS
The Edinburgh Building, Cambridge CB2 2RU, UK
40 West 20th Street, New York, NY 10011–4211, USA
10 Stamford Road, Oakleigh, VIC 3166, Australia
Ruiz de Alarcón 13, 28014 Madrid, Spain
Dock House, The Waterfront, Cape Town 8001, South Africa

http://www.cambridge.org

First published 2000

Printed in the United Kingdom at the University Press, Cambridge

Typeface Monotype Times 10/12pt *System* LaTeX 2_ε [UPH]

A catalogue record for this book is available from the British Library

Library of Congress Cataloguing in Publication data

Andrews, David G.
An introduction to atmospheric physics / David G. Andrews.
p. cm.
Includes index.
ISBN 0 521 62051 1 (hb). – ISBN 0 521 62958 6 (pbk)
1. Atmospheric physics. I. Title.
QC863.A596 2000
551.51–dc21 99-20191 CIP

ISBN 0 521 62051 1 hardback
ISBN 0 521 62958 6 paperback

Contents

Preface

Atmospheric physics has a long history as a serious scientific discipline, extending back at least as far as the late seventeenth century. Today it is a rich and fascinating subject, sustained by detailed global observations and underpinned by solid theoretical foundations. It provides an essential tool for tackling a wide range of environmental questions, on local, regional and global scales. Although the solutions to vital and challenging problems concerning weather forecasting and climate prediction rely heavily on the use of supercomputers, they rely even more on the imaginative application of soundly based physical insights.

This book is intended as an introductory working text for third- or fourth-year undergraduates studying atmospheric physics as part of a physics, meteorology or environmental science degree course. It should also be useful for graduate students who are studying atmospheric physics for the first time and for students of applied mathematics, physical chemistry and engineering who have an interest in the atmosphere.

Modern scientific study of the atmosphere draws on many branches of physics. I believe that a balanced introductory course in atmospheric physics should include at least some atmospheric thermodynamics, radiative transfer, atmospheric fluid dynamics and elementary atmospheric chemistry. Armed with some understanding of these topics, the interested student will be able to grasp the essential physics behind important issues of current concern, such as the amplification of the greenhouse effect and associated questions of climatic change, the Antarctic ozone hole and global depletion of ozone, as well as more familiar processes such as the formation of raindrops and the development of weather systems.

This book therefore aims to show how basic physical principles can be applied to help us to understand the workings of the Earth's atmosphere. It includes treatments of the topics mentioned in the previous paragraph, plus a few others. (However, for reasons of space I have omitted the beautiful, but less practically useful, subject of atmospheric optics.) Attention is restricted to the troposphere, stratosphere and mesosphere, that is, the region between the ground and about 90 km altitude. Although other planets are seldom mentioned explicitly, many of the topics covered also apply to the atmospheres of Venus and Mars.

The emphasis in the text is on the *underlying physics*, atmospheric applications being developed mainly in the problems given at the end of each chapter. It is essential that the serious student should attempt some of these problems, to test his or her understanding of the material and to obtain a broader perspective on the subject than can be provided by the text alone. In some cases important meteorological applications have been omitted because they rely on semi-empirical rules rather than on basic physics; there are excellent meteorology books covering this type of material.

The book assumes a basic knowledge of thermodynamics, electromagnetic radiation and quantum physics, together with some elementary vector calculus, at about the level reached in core physics courses at universities. It does not assume prior knowledge of fluid dynamics, which is frequently omitted from core physics courses[†]. Most of the material included here is based on a number of years' experience teaching atmospheric physics to undergraduate physicists at Oxford University.

Many colleagues have assisted me while I was writing this book or have provided constructive criticism of the text. I would particularly like to acknowledge help from Anu Dudhia, Brian Gardiner, Simon Hall, Alan Iwi, David Lary, Bryan Lawrence, Judith Lean, Stephen Lewis, Dingmin Li, Julius London, Paul Morris, Suzanne Rosier, Keith Shine, John Stanford, Adam Stephen, Fred Taylor, Geraint Vaughan and Bob Wells. Finally, I wish to express enormous thanks to my wife, Kathleen, for all her support and encouragement during the many months of planning and writing.

The following publishers have kindly given permission to reproduce or adapt figures:

The American Meteorological Society: Figure 1.7.
Kluwer Academic Publishers: Figures 3.13, 3.15 and 3.17.
Oxford University Press: Figures 5.1 and 5.2.
The Royal Meteorological Society: Figure 6.6.
The Optical Society of America: Figure 7.8.
Cambridge University Press: Figure 7.16.
Taylor and Francis Ltd: Figure 8.4.

I also thank the authors of these figures for permission to reproduce or adapt their material.

David Andrews
`http://www.atm.ox.ac.uk/user/andrews`

[†] In my view this omission is a mistake, since fluid dynamics provides a physical understanding of a host of readily observable – and scientifically and technologically important – phenomena, including many of those covered in this book.

CHAPTER ONE

Introduction

This chapter gives a quick sketch of some of the material to be covered in this book. We start in Section 1.1 with an outline of some of the more important physical processes that occur in the Earth's atmosphere. To interpret atmospheric observations we need to develop physical and mathematical models; they are briefly discussed in Section 1.2. In Section 1.3 two simple models are introduced; the second of these is a very basic representation of the greenhouse effect, which can be adapted to give some insight into aspects of global warming. In Section 1.4 we present a selection of observations of atmospheric processes, together with simple physical explanations for some of them. In Section 1.5 we briefly mention some ideas on weather and climate.

1.1 The atmosphere as a physical system

The Earth's atmosphere is a natural laboratory, in which a wide variety of physical processes takes place. The purpose of this book is to show how basic physical principles can help us model, interpret and predict some of these processes. This section presents a brief overview of the physics involved.

The atmosphere consists of a mixture of ideal gases: although molecular nitrogen and molecular oxygen predominate by volume, the minor constituents carbon dioxide, ozone and water vapour play crucial roles. The forcing of the atmosphere is primarily from the Sun, though interactions with the land and the ocean are also important.

The atmosphere is continually bombarded by solar photons at infra-red, visible and ultra-violet wavelengths. Some solar photons are scattered back to space by atmospheric gases or reflected back to space by clouds or the Earth's surface; some are absorbed by atmospheric molecules (especially water vapour and ozone) or clouds, leading to heating of parts of the atmosphere; and some reach the Earth's surface and heat it. Atmospheric gases (especially carbon dioxide, water vapour and ozone), clouds and the Earth's surface also emit and absorb infra-red photons, leading to further heat transfer between one region and another, or loss of heat to space. Some of these *radiative-transfer* processes are discussed in Chapter 3. Solar photons may also be energetic enough to disrupt molecular chemical bonds, leading to photochemical reactions; see Chapter 6.

In general, each portion of the atmosphere is approximately in hydrostatic equilibrium†; that is, its weight is supported by the difference in pressure between its lower and upper surfaces. An alternative statement of this physical fact is that there is a balance between vertical pressure gradients and the gravitational force per unit volume acting on each portion of the atmosphere. On combining the equation describing hydrostatic balance with the ideal gas law we find that, in a hypothetical *isothermal* atmosphere, the pressure and density would fall exponentially with altitude (see Section 2.3). In the real, non-isothermal, atmosphere the pressure and density variations are usually still close to this exponential form, with an *e*-folding height of about 7 or 8 km. Gravity thus tends to produce a density *stratification* in the atmosphere.

Given a density stratification of this kind, a small portion of air that is displaced upwards from its equilibrium position will be negatively buoyant compared with its surroundings and hence will fall back towards equilibrium, under gravity; similarly a downward-displaced portion will rise back towards its equilibrium position. Buoyancy therefore acts as a restoring force; the atmosphere is said to be *stably stratified.* The strength of the stability of the stratification varies from one part of the atmosphere to another.

Thermodynamic principles are essential for describing many atmospheric processes. For example, any consideration of the effects of atmospheric heating or cooling will make use of the First Law of Thermodynamics. The concept of entropy (or the closely related quantity, potential temperature) frequently assists interpretation of atmospheric behaviour. Knowledge of changes in phase between vapour, liquid and solid forms of the water in the atmosphere is crucial for an understanding of the formation of rain and snow. Moreover, the associated latent heating and cooling can provide important contributions to heat transfer within the atmosphere–ocean system; for example, evaporation of a droplet of sea-water at one location and subsequent condensation of the resulting water vapour into a droplet at another location in the atmosphere transfers heat from the ocean to the atmosphere. The basics of atmospheric thermodynamics are covered in Chapter 2.

† This is usually valid on horizontal scales larger than a few kilometres.

In atmospheric physics we use the usual macroscopic definitions of the temperature and pressure of a gas. From the kinetic theory of gases, these have well-known interpretations in terms of the mean kinetic energy of molecules and the mean transfer of momentum by molecules, respectively. When considering dynamical processes – that is, the response of atmospheric motions to applied forces – we can average other physical quantities such as density and velocity over many molecules and regard the atmosphere as a continuous fluid; individual molecular motions need not be taken into account. It is clear from the most cursory weather observations that the resulting bulk fluid motion of the atmosphere is still very complex. However, when the motion is viewed on a large scale (say hundreds of kilometres in the horizontal direction), some simplifying features appear. In particular, Coriolis forces play significant roles: these forces result from the rotation of the Earth and tend to deflect a moving portion of air to the right of its motion in the Northern Hemisphere and to the left in the Southern Hemisphere. A near balance between Coriolis forces and horizontal pressure gradient forces leads to wind motions that circulate along isobars (surfaces of constant pressure) at a given height. The sense of the circulation is anticlockwise around low-pressure regions and clockwise around high-pressure regions in the Northern Hemisphere and vice versa in the Southern Hemisphere. The basic principles of atmospheric fluid dynamics are introduced in Chapter 4.

An important feature of the buoyancy restoring effect in a stably stratified atmosphere is that it can support fluid-dynamical waves, known as *gravity waves*†, in which the fluid pressure, density, temperature and velocity fluctuate together. These waves may propagate, allowing one part of the atmosphere to 'communicate' over great distances with other parts, without a corresponding transport of mass. The Coriolis force can also act as a restoring force, giving rise to further types of fluid wave motion. In particular, on large scales we find *Rossby waves*, which depend crucially on the rotation and the spherical geometry of the Earth and are associated with many observed large-scale disturbances in the troposphere and the stratosphere. As in many other branches of physics, the study of wave motions is an essential part of atmospheric physics; see Chapter 5.

As noted above, solar radiation can initiate photochemical reactions by dissociating atmospheric molecules. A host of other types of chemical reaction between atmospheric molecules, both natural and man-made, can also occur. Atmospheric chemistry is a large and highly complex subject; in this book we focus on one small but significant branch of the subject, namely the chemistry of stratospheric ozone. This provides a good example of physical principles

† Not to be confused with the gravitational waves of General Relativity!

in operation and is highly topical, with direct application to the Antarctic ozone hole and global depletion of ozone; see Chapter 6. It also demonstrates the importance of atmospheric transport processes, by which the winds blow chemical species from one part of the globe to another.

No study of the atmosphere can make progress without suitable observations and all observational techniques rely to some extent on physical principles. One important type of observational technique is that of *remote sounding*, which depends on the detection of electromagnetic radiation emitted, scattered or transmitted by the atmosphere. In Chapter 7 we describe several examples of remote sounding, looking both at space-borne and at ground-based methods.

1.2 Atmospheric models

Unlike laboratory physicists, atmospheric researchers cannot perform controlled experiments on the large-scale atmosphere. The standard 'scientific method', of observing phenomena, formulating hypotheses, testing them by experiment, then formulating revised hypotheses and so on, cannot be applied directly. Instead, after an atmospheric phenomenon is discovered, perhaps by sifting through a great deal of data, we develop *models*, which incorporate a selection of processes that we hypothesise are significant for the phenomenon. Models act as surrogate atmospheres, on which 'thought experiments' *can* be performed. These models are usually formulated in terms of mathematical equations and the 'experiments' are performed by solving these equations (perhaps by computer) under various conditions and interpreting the solutions in terms of physical behaviour. Occasionally a laboratory apparatus may provide a useful atmospheric model. The performance of the model (and thus the appropriateness of the hypothesised set of processes) is judged by comparing the model's behaviour with that of the atmosphere.

Normally, a *hierarchy* of models is used, starting with simple 'back-of-the-envelope' models and progressing on to the highly complex 'general circulation models' which require large computer resources. The models considered in this book are mostly of the simpler type, although the more complex ones are briefly discussed in Chapter 8. Since the simpler models can usually be investigated analytically and their workings fully explored, they can provide a basic 'physical intuition', which can then be applied to the interpretation of the more complex models. Because of their very simplicity, however, they cannot usually be expected to give accurate simulations of observed atmospheric behaviour.

The more comprehensive models bring together many of the physical principles introduced in this book and allow for interac-

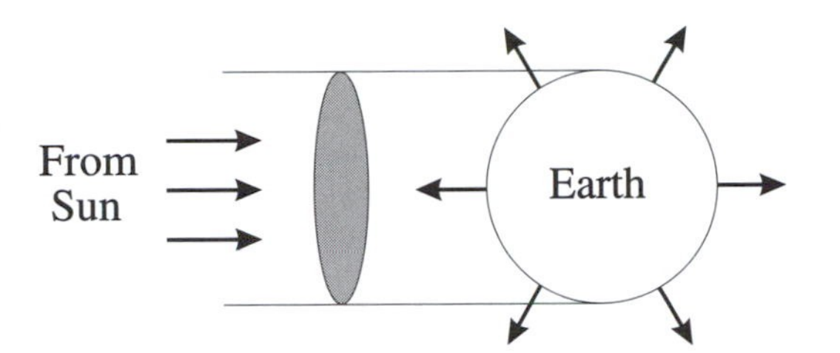

Figure 1.1 Illustrating the calculation of the temperature of the Earth, ignoring any trapping of radiation by the atmosphere. The parallel arrows indicate solar radiation, confined within a tube of cross-sectional area πa^2. The radial arrows indicate outgoing thermal radiation from the total surface area $4\pi a^2$ of the Earth.

tions between them. Some of these interactions may involve complex feedbacks, however, so it may be difficult to establish causal relationships among the various processes involved.

Historically, a major use of comprehensive atmospheric models has been for weather forecasting. With the development of supercomputers, reliable longer-term climate forecasting is becoming a feasible proposition, although the models used for this purpose are still deficient in many respects. Complex models are also used for *data assimilation*, by providing a dynamically self-consistent means of interpolating, in space and time, between sparse data points from a variety of sources. Data assimilation is nowadays a vital component of the weather-forecasting process.

1.3 Two simple atmospheric models

It is a basic observational fact that the Earth's mean surface temperature is about 288 K. In this section we consider whether this can be explained in simple terms, given the input of solar radiation and some elementary atmospheric physics. We consider two models; the first turns out to be seriously defective, but the second, which includes a simple representation of the *greenhouse effect*, gives a surface temperature in reasonable agreement with observations. The second model also gives us a very simple way of starting to quantify the effects of the processes that may lead to global warming.

1.3.1 A simple radiative model

The incident flux, or power per unit area, of solar energy at the Earth's mean distance from the Sun (the so-called *solar constant*) is $F_s = 1370\,\mathrm{W\,m^{-2}}$. The solar beam is essentially parallel at the Earth, so the power that is intercepted by the Earth is contained in a tube of cross-sectional area πa^2, where a is the Earth's radius; see figure 1.1. The total solar energy received per unit time is therefore $F_s \pi a^2$.

We assume that the Earth–atmosphere system has an *albedo* A equal to 0.3; that is, 30% of the incoming solar radiation is reflected back to space without being absorbed: this is close to the observed value. The Earth therefore reflects $0.3F_s\pi a^2$ of the incoming solar power back to space.

If the Earth is assumed to emit as a *black body* at a uniform absolute temperature T then, by the Stefan–Boltzmann law,

$$\text{Power emitted per unit area} = \sigma T^4, \tag{1.1}$$

where σ is the Stefan–Boltzmann constant†. However, power is emitted in all directions from a total surface area $4\pi a^2$, so the total power emitted is $4\pi a^2\sigma T^4$. We assume in the present model that all of this power is transmitted to space, with none absorbed by the atmosphere. Then, assuming that the Earth is in thermal equilibrium, the incoming and outgoing power must balance, so

$$(1 - A)F_s\pi a^2 = 4\pi a^2\sigma T^4.$$

On substituting the values of A and F_s into this, we obtain $T \approx$ 255 K. This value is significantly lower than the observed mean surface temperature of about 288 K. The present model is clearly lacking in some vital ingredient; we find in Section 1.3.2 that inclusion of the radiation-trapping effect of the atmosphere (the 'greenhouse effect') leads to a surface temperature that is much closer to reality.

1.3.2 A simple model of the greenhouse effect

We now consider the effect of adding a shallow layer of atmosphere, of uniform temperature T_a, to the model of Section 1.3.1; see figure 1.2. The atmosphere is assumed to transmit a fraction $\mathcal{T}_s$ of any incident solar (short-wave) radiation and a fraction $\mathcal{T}_t$ of any incident thermal (infra-red, or long-wave) radiation‡, and to absorb the remainder. We assume that the ground is at the temperature T_g.

Taking account of albedo effects and the difference between the area of the emitting surface $4\pi a^2$ and the intercepted cross-sectional area πa^2 of the solar beam (see Section 1.3.1), the mean unreflected incoming solar flux (or irradiance: see Section 3.2.1) at the top of the atmosphere is

$$F_0 = \tfrac{1}{4}(1 - A)F_s, \tag{1.2}$$

or about 240 W m^{-2}. Of this, a proportion $\mathcal{T}_sF_0$ reaches the ground, the remainder being absorbed by the atmosphere.

† The concept of a black body is explained in Section 3.1.1; for the moment, all that is required is that the power per unit area emitted by a black body satisfies equation (1.1). The value of σ is given in Appendix A, together with the values of other useful physical constants.

‡ These fractions are called *transmittances*; see Section 3.4.

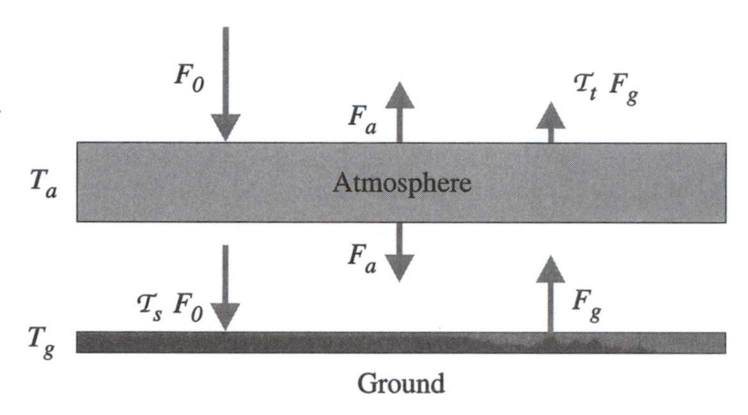

Figure 1.2 A simple model of the greenhouse effect. The atmosphere is taken to be a shallow layer at temperature T_a and the ground a black body at temperature T_g. Various solar and thermal fluxes are shown.

The ground is assumed to emit as a black body. By equation (1.1) it therefore emits an upward flux $F_g = \sigma T_g^4$, of which a proportion $\mathcal{T}_t F_g$ reaches the top of the atmosphere, the remainder being absorbed by the atmosphere. The atmosphere is not a black body, but emits fluxes $F_a = (1 - \mathcal{T}_t)\sigma T_a^4$ both upwards and downwards, as shown in Figure 1.2†.

We can now balance these fluxes, assuming that the system is in equilibrium. At the top of the atmosphere we have

$$F_0 = F_a + \mathcal{T}_t F_g, \tag{1.3a}$$

and at the ground

$$F_g = F_a + \mathcal{T}_s F_0. \tag{1.3b}$$

By eliminating F_a from equations (1.3) we obtain

$$F_g = \sigma T_g^4 = F_0 \frac{1 + \mathcal{T}_s}{1 + \mathcal{T}_t}. \tag{1.4}$$

In the absence of an absorbing atmosphere, we would have $\mathcal{T}_s = \mathcal{T}_t = 1$, so F_g would equal F_0, giving $T_s \approx 255\,\mathrm{K}$, as in Section 1.3.1. However, rough values for the Earth's atmosphere are $\mathcal{T}_s \approx 0.9$ (strong transmission and weak absorption of solar radiation) and $\mathcal{T}_t \approx 0.2$ (weak transmission and strong absorption of thermal radiation), so that $F_g \approx 1.6 F_0$, leading to a surface temperature of $T_g \approx 1.6^{1/4} \times 255 \approx 286\,\mathrm{K}$, which is quite close to the observed mean value of 288 K. (The close agreement is partly fortuitous, however, since in reality non-radiative processes also contribute to the energy balance.)

This is a simple example of the greenhouse effect: the greater temperature depends on the fact that there is less absorption (greater transmission) for solar radiation than there is for thermal radiation.

† By Kirchhoff's Law, the emittance – the ratio of the actual emitted flux to the flux that would be emitted by a black body at the same temperature – equals the absorptance $1 - \mathcal{T}_t$; see Section 3.1.1.

Thus the atmosphere readily transmits solar radiation but tends to trap thermal radiation†. We can also find the atmospheric flux from equations (1.3):

$$F_a = (1 - \mathcal{T}_t)\sigma T_a^4 = F_0 \frac{1 - \mathcal{T}_s \mathcal{T}_t}{1 + \mathcal{T}_t}$$

and this gives the temperature of the shallow atmosphere in this model, $T_a \approx 245\,\mathrm{K}$.

1.3.3 Global warming

The model described in Section 1.3.2 provides an indication of the way in which the greenhouse effect increases the Earth's mean surface temperature by some 30 K above the value that would occur in the absence of a radiation-trapping atmosphere. To understand mechanisms of potential global warming we must go further and investigate the *enhanced greenhouse effect*, namely the response of the surface temperature to *perturbations* in the concentration of radiatively active gases such as carbon dioxide. This is of course an enormously complex question, but a very simple start can be made by using equations (1.2) and (1.4) to find how T_g changes in response to small increments $\Delta\mathcal{T}_s$ in the transmission of solar radiation, $\Delta\mathcal{T}_t$ in the transmission of thermal radiation and ΔA in the albedo, with a fixed solar constant F_s. By taking logarithms and differentiating, we find that the increment ΔT_g in surface temperature is given by

$$4\frac{\Delta T_g}{T_g} = \frac{\Delta\mathcal{T}_s}{1 + \mathcal{T}_s} - \frac{\Delta\mathcal{T}_t}{1 + \mathcal{T}_t} - \frac{\Delta A}{1 - A}.$$

If the concentration of a radiatively active gas increases, this will generally lead to decreases in the solar and thermal transmittances; however, detailed consideration may be required in order to be able to infer whether the albedo increases or decreases. A decrease in the thermal transmittance ($\Delta\mathcal{T}_t < 0$) implies more trapping of thermal radiation and hence contributes to an increase in T_g. On the other hand, a decrease in the solar transmittance means that less solar radiation reaches the ground and hence contributes to a decrease in T_g. An increase in albedo means that more solar radiation is reflected back to space, contributing to a decrease in T_g. Which process dominates depends on the detailed radiative properties of the gas in question. A more realistic model of the greenhouse effect is given in Section 3.7.

† The term 'greenhouse effect' is a misnomer, however, since the elevated temperature in a greenhouse does not primarily depend on the similar radiative properties of glass, but rather on the suppression of convective heat loss.

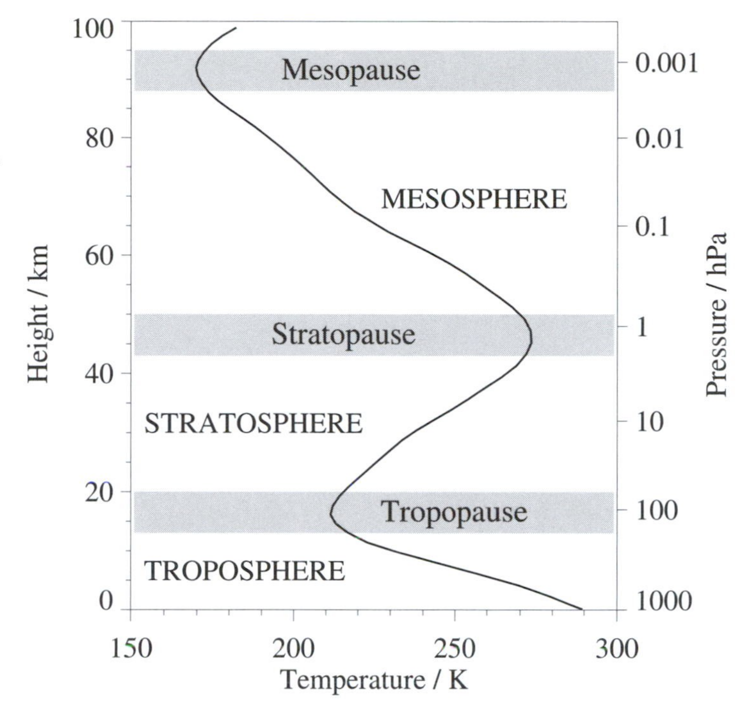

Figure 1.3 Typical vertical structure of atmospheric temperature (K) in the lowest 100 km of the atmosphere. Based on data from Fleming *et al.* (1990).

1.4 Some atmospheric observations

In this section we present a selection of examples of basic atmospheric observations and give some indication of their physical explanation. Further details are given in later chapters of this book.

1.4.1 The mean temperature and wind fields

Figure 1.3 shows a typical example of the vertical structure of the temperature in the lowest 100 km of the atmosphere. The atmosphere is conventionally divided into layers in the vertical direction, according to the variation of temperature with height. The layer from the ground up to about 15 km altitude, in which the temperature decreases with height, is called the *troposphere* and is bounded above by the *tropopause*. The layer from the tropopause to about 50 km altitude, in which the temperature rises with altitude, is called the *stratosphere* and is bounded above by the *stratopause*. The layer from the stratopause to about 85–90 km, in which the temperature again falls with altitude, is called the *mesosphere* and is bounded above by the *mesopause*. Above the mesopause is the *thermosphere*, in which the temperature again rises with altitude.

The troposphere is also called the *lower atmosphere*. It is here

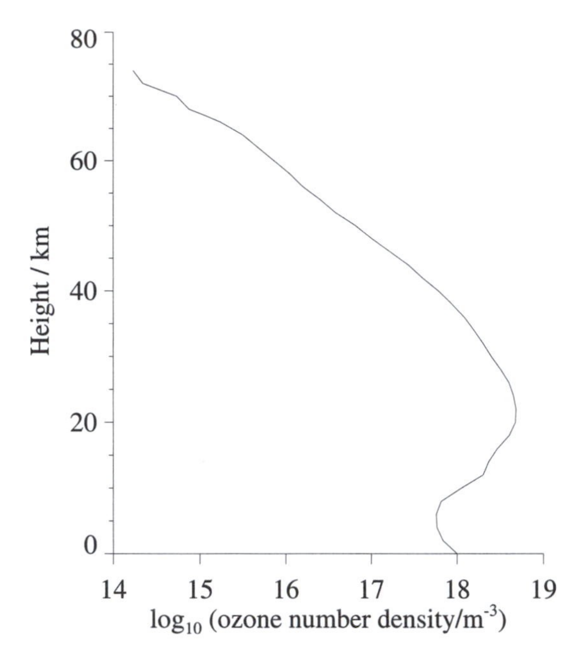

Figure 1.4 Typical vertical structure of the mean midlatitude ozone number density (molecules m^{-3}). Based on data from U. S. Standard Atmosphere (1976).

that most 'weather' phenomena, such as cyclones, fronts, hurricanes, rain, snow, thunder and lightning, occur.

The stratosphere and mesosphere together are called the *middle atmosphere*. A notable feature of the stratosphere is that it contains the bulk of the ozone molecules in the atmosphere; see Figure 1.4. The neighbourhood of the ozone maximum in the lower stratosphere is loosely known as the *ozone layer*. The production of ozone (O_3) molecules occurs through photochemical processes involving the absorption of solar ultra-violet photons by molecular oxygen (O_2) in the stratosphere, three O_2 molecules eventually forming two O_3 molecules. The equilibrium profile of ozone depends also on chemical ozone-destruction processes and on the transport of ozone by the winds (see Chapter 6).

Above the middle atmosphere is the *upper atmosphere*, where effects of ionisation become dominant in determining the atmospheric structure and the air becomes so rarefied that the assumption that it can be treated as a continuous fluid starts to break down. In this book we concentrate on the physics of the lower and middle atmospheres.

From hydrostatic balance, the pressure at any level in the atmosphere is proportional to the mass of air above that level. From the pressure axis in Figure 1.3 it follows that approximately 90% of the atmospheric mass is in the troposphere, a little under 10% in the stratosphere and only about 0.1% in the mesosphere and above†.

† In this book we use the unit hPa for pressure (1 hPa = 10^2 Pa). This is equivalent to the millibar, which is commonly used in meteorology.

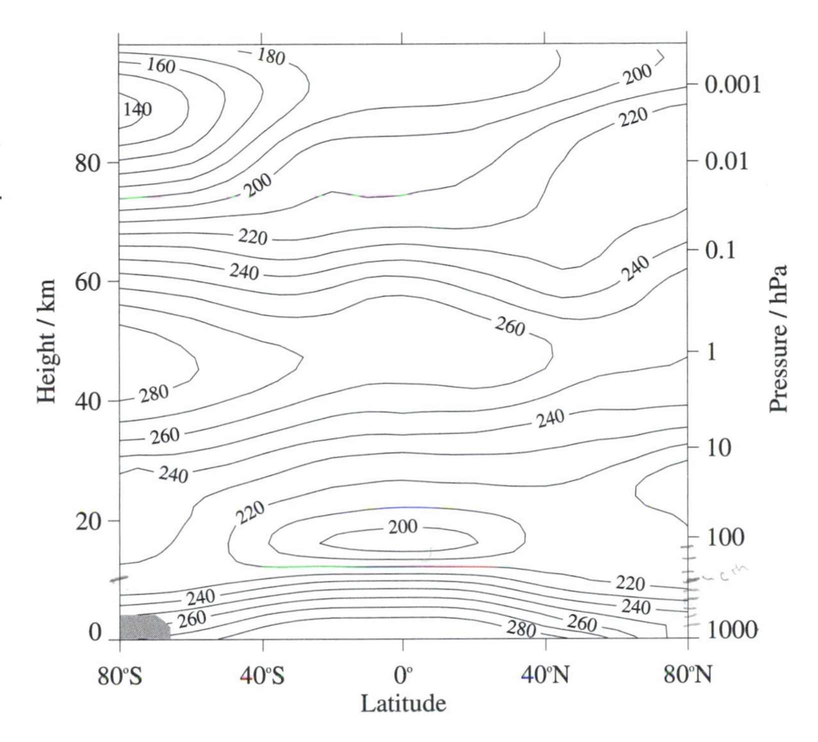

Figure 1.5 Zonal-mean temperature (K) for January, from the CIRA dataset. A small region at low levels over Antarctica is omitted. Based on data from Fleming *et al.* (1990).

Despite their relatively low mass, the stratosphere and mesosphere are not insignificant, however. For example, ozone in the stratosphere absorbs ultra-violet solar radiation, thereby protecting the biosphere from potentially damaging effects.

Figure 1.3 is not representative of all latitudes and seasons. A more comprehensive plot, of the zonal-mean (i.e., longitudinally averaged) temperature averaged over several Januaries, as a function of height and latitude, is given in Figure 1.5. It will be seen that although the general shape of the vertical variation of temperature in midlatitudes is roughly in accord with that in Figure 1.3, there are significant latitudinal variations of the heights and magnitudes of the temperature extrema. For example, the equatorial tropopause is at a greater altitude and colder than that at higher latitudes, the summer stratopause is lower and warmer than the winter stratopause and the summer mesopause is extremely cold. (In fact the lowest natural terrestrial temperatures are found there.)

Some of these temperature features can be crudely explained in terms of simple physical mechanisms. For example, the warm stratopause can be attributed to the ozone distribution, which peaks in the stratosphere; absorption of solar radiation by the ozone leads to heating of the upper stratosphere and, since in equilibrium this heating is balanced mainly by infra-red cooling, there must be a

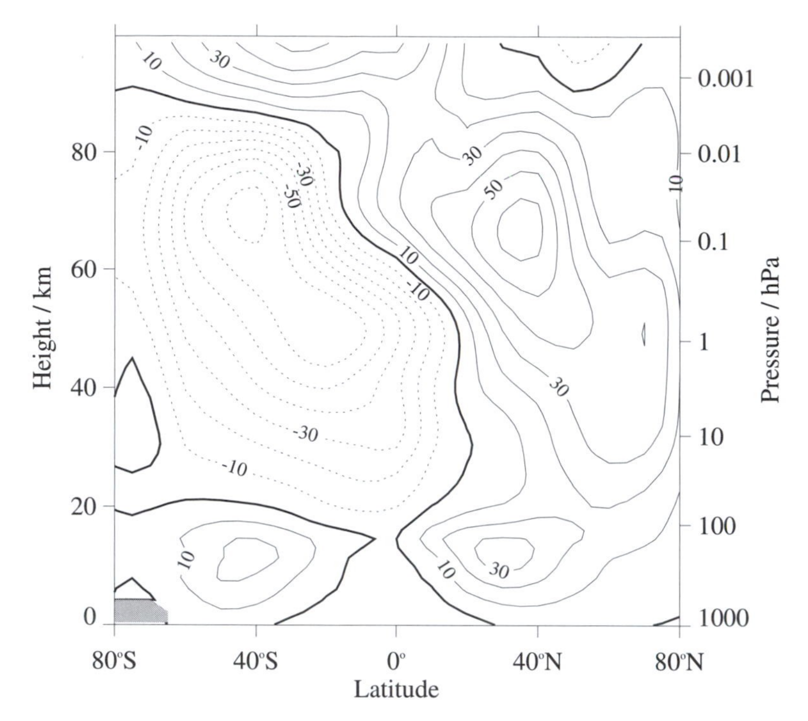

Figure 1.6 Zonal-mean zonal wind ($m\,s^{-1}$) for January, from the CIRA dataset. Thin solid lines: eastward winds; thick solid line: zero winds; dashed lines: westward winds. A small region at low levels over Antarctica is omitted. Based on data from Fleming *et al.* (1990).

local temperature maximum, so that heat can radiate to cooler regions.

If this radiative control of temperature were to continue down to the ground, the temperature in the troposphere would decrease much more rapidly with height than is observed and this temperature profile (and its associated density profile) would be statically unstable. Such a temperature profile could not persist, but might be expected to give rise to convective activity that would modify the temperature profile, causing it to decrease less rapidly with height until it was just statically stable again. This process appears to occur in the moist tropical troposphere, where the temperature decrease with altitude is fairly close to the *saturated adiabatic lapse rate*, the rate of decrease with height of the temperature of a parcel of air saturated with water vapour, which condenses (releasing latent heat) as the parcel rises; see Section 2.8. At higher latitudes other processes, such as midlatitude cyclones and anticyclones, may also play a part in determining the temperature profile. The position of the tropopause depends on an interplay between the processes that lead the temperature to fall with height in the troposphere and increase with height in the stratosphere; the precise details are a subject of active current research.

The explanations of the cold equatorial tropopause and extremely cold summer mesopause are quite complex. It turns out that there is

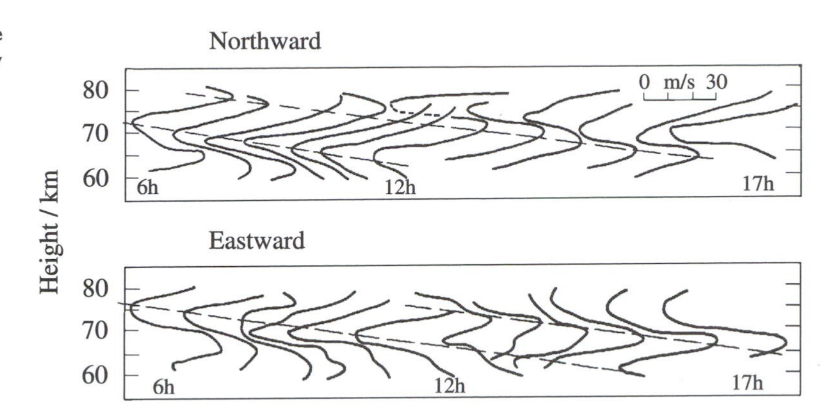

Figure 1.7 An example of an atmospheric gravity wave over Alaska, as measured by a ground-based radar. The northward and eastward wind components (in $\mathrm{m\,s^{-1}}$) are shown as a function of height, between altitudes of 60 and 80 km, and at each hour over a period of about 17 hours. From Balsley *et al.* (1983).

dynamically driven rising motion in both of these regions; the rising air expands as it moves to lower pressure and cools. Conversely there is descent over the winter pole, leading to temperatures that are warmer than would otherwise occur.

The zonal-mean zonal (west-to-east) winds for January are shown in Figure 1.6. These are related to the temperature field in Figure 1.5 by *thermal windshear balance*, as expressed by equation (4.28b). In the troposphere, the mean zonal winds are generally eastward at midlatitudes, with two prominent 'jet streams', and westward at low latitudes†. In the stratosphere and mesosphere the mean zonal winds are generally eastward in winter and westward in summer. The winds at low latitudes shown in Figure 1.6 are not representative of every January, partly because they do not follow a simple annual cycle in the equatorial lower stratosphere; there is a prominent *quasi-biennial oscillation* there, with a period of approximately 28 months.

1.4.2 Gravity waves

Figure 1.7 shows northward and eastward wind components between altitudes of 60 and 80 km, over Alaska, measured by a ground-based radar. The radar transmits radio waves almost vertically and measures the backscattered signal; from this, wind velocities can be determined; see Section 7.3.2. Roughly sinusoidal wind fluctuations in the vertical are seen, with a wavelength of about 15 km and propagating downwards in time with a period of about 9 h.

These quasi-sinusoidal fluctuations are strongly suggestive of some kind of wave motion and further study confirms this idea. The waves are an example of the fluid-dynamical gravity waves

† The terms 'westerly', meaning eastward, and 'easterly', meaning westward, are commonly used in meteorology, but will be avoided in this book.

mentioned in Section 1.1. Atmospheric gravity waves are analogous to horizontally propagating surface waves on water, which depend on the restoring mechanism provided by the contrast in density between air and water: the water in a wave crest is denser than the surrounding air and tends to fall, while the air in a wave trough is lighter than the surrounding water and tends to rise. We can imagine the smooth vertical density variation in the atmosphere to be approximated by a stack of thin fluid layers, whose densities decrease with height. The surface-wave mechanism operates at each density interface, so now the wave can propagate vertically as well as horizontally.

The particular type of gravity wave shown in Figure 1.7 is called an inertia–gravity wave (see Section 5.4); it is actually of large enough horizontal scale (a few hundred kilometres) and period to be influenced to some extent by the Earth's rotation. These measurements provide an unusually clear example of a sinusoidal oscillation: in most cases, too many other dynamical processes are occurring in the atmosphere for waves to be very easily identified and careful data analysis must be performed to isolate them.

Perhaps surprisingly, the downward phase progression of the waves in Figure 1.7 indicates *upward* propagation of 'information' by the waves (i.e. an upward group velocity). Gravity waves will be studied in detail in Section 5.4 and, among other things, it will be shown that this type of wave is *dispersive*; the phase and group velocities can therefore be in different directions. This is just one of the ways in which the propagation characteristics of the atmospheric waves studied in this book differ from those of the more familiar non-dispersive waves such as electromagnetic waves in a vacuum.

Gravity waves are generated in many different ways, including by air flow over mountains and by convective activity in the troposphere. Waves generated in the lower atmosphere may propagate upwards into the stratosphere and mesosphere. As the background air density decreases, the amplitudes of the wave fluctuations in wind (and associated fluctuations in temperature and density) will rise. As a result, gravity waves may attain large amplitudes in the mesosphere and exert a considerable influence on the mean atmospheric state there.

1.4.3 Rossby waves

Figure 1.8 depicts the temperature in the Northern Hemisphere, at a level near 24 km altitude, during a period when the stratosphere was disturbed by a vigorous dynamical event known as a *stratospheric warming*. A cold region is located on one side of the pole (roughly along 0° E) and a warm region on the other (roughly

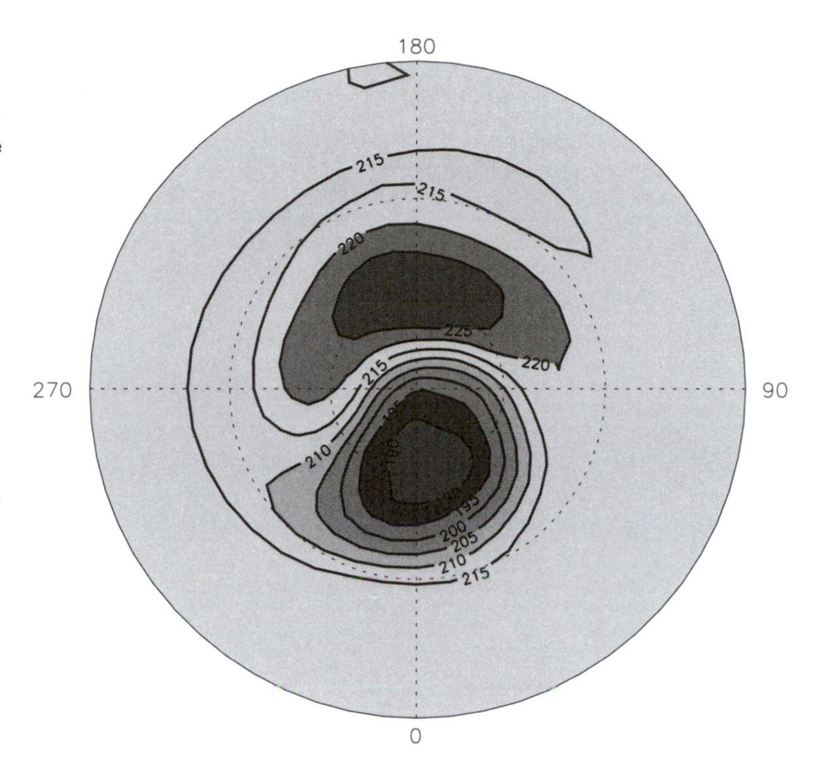

Figure 1.8 Polar stereographic map of atmospheric temperature (K) near 24 km altitude in the Northern Hemisphere stratosphere on 9 January 1992, as measured by the Improved Stratospheric and Mesospheric Sounder (ISAMS) on the Upper Atmospheric Research Satellite (UARS). The North Pole is at the centre, the 60° N and 30° N latitude circles are shown and the equator is the outer circle; four longitudes are also shown. (Diagram supplied by Dr A. M. Iwi.)

along 180° E). Moving around a latitude circle near the pole, we find that the temperature varies roughly sinusoidally with longitude, one wavelength encompassing 360° of longitude. The fact that the cold part of the disturbance is smaller and stronger than the warm part shows that the fluctuation is not exactly sinusoidal. However, the phenomenon is wave-like in many respects and is an example of the Rossby wave mentioned in Section 1.1. The horizontal wavelength is several thousand kilometres in extent and the wave's dynamics are quite different from those of the gravity wave. The propagation mechanism is quite subtle, depending both on the rotation and on the curvature of the Earth; the details are given in Section 5.5. Rossby waves, like gravity waves, are dispersive.

Several types of Rossby wave are observed in the atmosphere. Some are *stationary*, that is their wave patterns are fixed with respect to the Earth; since they are dispersive they may still propagate information, because the group velocity may be non-zero even if the phase velocity is zero. Others have patterns that travel with respect to the Earth. Strong Rossby-wave disturbances in the stratosphere, such as that shown in Figure 1.8, are frequently due to upward propagation of waves generated by large-scale weather disturbances in the troposphere. It is found that, when the background winds are eastward, as in winter (see Figure 1.6), only the longest-wavelength stationary Rossby waves can propagate vertically. This accounts for

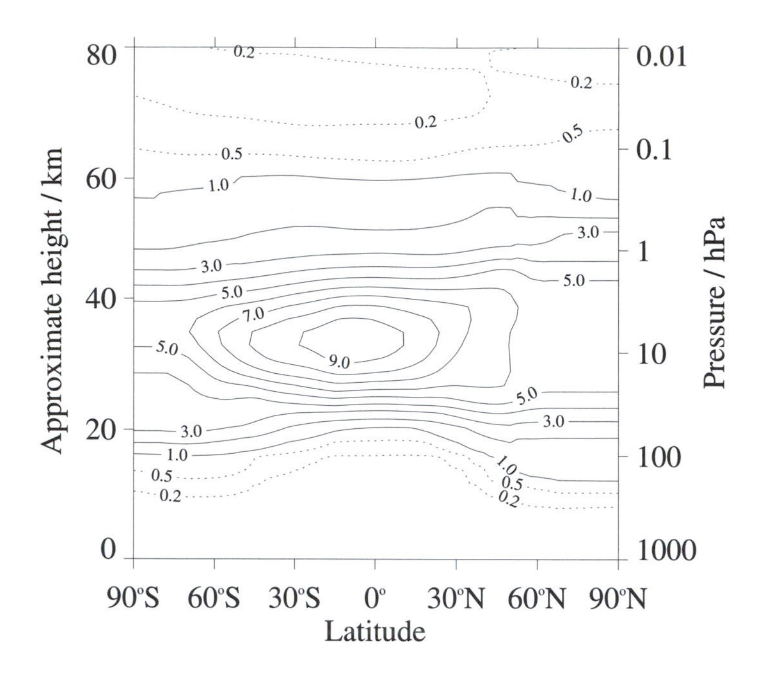

Figure 1.9 Zonal-mean volume mixing ratio of ozone (parts per million by volume), as a function of latitude and height, for January, based on the 5-year climatology of Li and Shine (1995). Data provided by Dr D. Li.

the observed prevalence of larger-scale disturbances in the winter stratosphere.

1.4.4 Ozone

As mentioned in Section 1.4.1, atmospheric ozone is important because it absorbs solar ultra-violet radiation, thus protecting human and animal life from potentially harmful consequences. It is therefore crucial to have good measurements of ozone concentrations and a good understanding of the processes by which it is produced and destroyed.

A typical vertical profile of the ozone number density was shown in Figure 1.4. An alternative measure of the concentration of a gaseous constituent of air, such as ozone, is the *volume mixing ratio*, that is, the number density of the constituent divided by the total number density of the 'air'. A useful property of the volume mixing ratio is that, unlike the number density, it is constant for a moving parcel of air in the absence of chemical production and loss processes. A latitude–height cross-section of the zonal-mean volume mixing ratio of ozone for January is given in Figure 1.9. This shows that the maximum values are in the low-latitude stratosphere (where the photochemical production of ozone is greatest) but also that there are significant values over the winter pole, where no production takes place since the Sun is below the horizon all day.

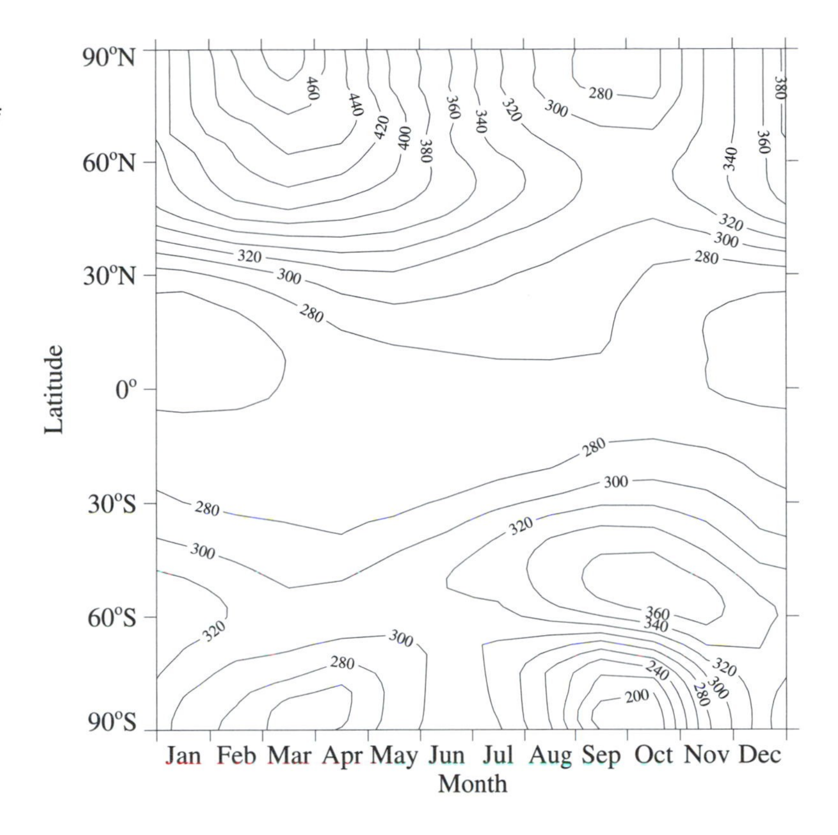

Figure 1.10 The observed annual cycle in column ozone, based on the 5-year climatology of Li and Shine (1995). The units are Dobson Units: see Section 6.7. Data provided by Dr D. Li.

This suggests that ozone is transported into the 'polar night' region by wind motions.

Further information on the global ozone distribution is given in Figure 1.10, which shows the 'column ozone', a measure of the total number of ozone molecules in a vertical column of atmosphere, as a function of latitude and season. Low values of column ozone are found all year round at low latitudes, despite the fact that this is where most production of ozone takes place. Maximum amounts are found in spring: in the Northern Hemisphere this maximum occurs at high latitudes, whereas in the Southern Hemisphere it occurs at middle latitudes; a relative minimum is found in the Antarctic spring.

The Antarctic spring minimum of column ozone in Figure 1.10 is a manifestation of the *Antarctic ozone hole*, a dramatic reduction of ozone in the Antarctic stratosphere that has been observed in spring in recent years. Intensive international efforts, both observational and theoretical, have established that this loss of ozone is due to complex chemical and physical processes, involving chlorine resulting from man-made compounds such as chlorofluorocarbons (CFCs), taking place on the ice particles that can form in the extremely low Antarctic winter temperatures. The ozone lost during spring over Antarctica is mostly replenished there later in the year, but this effect may be contributing to a slow global loss of ozone. This research led to the Montréal Protocol, in which a phasing out of CFCs was agreed by the international community. Depletion of ozone is also occurring at other latitudes and is currently the subject of global monitoring and modelling. Further details are given in Chapter 6.

1.5 Weather and climate

The word *weather*, while having a clear enough meaning to the layperson, does not have a precise definition in atmospheric physics. However, it tends to encompass tropospheric events associated with atmospheric flows with length scales of hundreds of metres and more and time scales of a few days or less, although weather patterns may occasionally persist for a week or two. Weather phenomena are notoriously irregular†, but atmospheric data averaged over a month or so usually behave in a more regular manner. Most localities have regular longer-term variations in the average weather conditions, with a roughly annual cycle. However, this annual cycle does not repeat precisely from one year to the next: there may be dramatic interannual variations in the average weather at a given place.

Much of the thrust of atmospheric research has been directed towards the improvement of weather forecasting, for periods of up to two weeks ahead. Although present-day forecasting is partly limited by the insufficiency of computer power and incompleteness of atmospheric data, it is also limited to some extent by gaps in our understanding of the basic physics and dynamics of the atmosphere.

The word *climate* refers to the state of the atmosphere on longer time scales, typically averaged over several years or more. The understanding of climate and climate change does not necessarily require a complete understanding of every weather event; conversely there are physical processes, operating on long time scales, that are unimportant for weather prediction but crucial for climate prediction. (An example is heat transport from the deep ocean, which may vary on decadal or longer time scales.)

There is much current concern about whether human activity may be changing the climate significantly, principally through an amplification of the greenhouse effect due to increasing levels of carbon dioxide resulting from the burning of fossil fuels and the destruction of the tropical rain forests. Given the long time scales

† E. N. Lorenz's famous pioneering work on chaotic systems originated in a study of a simple nonlinear atmospheric model and was motivated by considerations of weather prediction; see Lorenz (1963).

involved, the detection of changes in climate is difficult; however, the latest estimates suggest, for example, that the global mean surface air temperature has increased by 0.3–0.6 °C over the last century. It is even more difficult to distinguish between man-made and natural climate change, given the complexity of the climate system (including the atmosphere, oceans, ice cover, biosphere and solid Earth). Since controlled experiments cannot be performed on the climate system, we must use models to identify cause-and-effect relationships, as noted in Section 1.2. Even the most sophisticated current models are still over-simplified. However, the current consensus among climate scientists, represented for example by the 1995 report of the Intergovernmental Panel on Climate Change (Houghton *et al.* 1996), is that the balance of evidence suggests that there is a discernible effect of human activity on the global climate.

References

More advanced treatments of some of the atmospheric physical processes discussed in this chapter are given in the texts by Houghton (1986), Goody (1995) and Salby (1996). The book by Wallace and Hobbs (1977) is an introductory text, with a sound physical basis and with meteorological applications in mind. A more descriptive account of weather and climate processes, requiring a lower level of physics, is given by McIlveen (1991). Holton (1992) provides an excellent introduction to atmospheric dynamics. A non-technical account of atmospheric waves is given by Andrews (1991). The physical principles of climate and climate change are expounded by Hartmann (1994).

CHAPTER TWO

Atmospheric thermodynamics

In this chapter we show how basic thermodynamic concepts can be applied to the atmosphere. We first note in Section 2.1 that the atmosphere behaves as an ideal gas. Some basic information on the various gases comprising the atmosphere is presented in Section 2.2. The fact that the atmosphere is fairly close to being in hydrostatic balance is used in Section 2.3 to develop some very simple ideas about the vertical structure of the atmosphere. An important quantity related to entropy, the *potential temperature*, is discussed in Section 2.4. The concept of an *air parcel* is introduced in Section 2.5 and is used to develop ideas about atmospheric stability and buoyancy oscillations. A brief introduction to the concept of *available potential energy* is given in Section 2.6.

The rest of the chapter is devoted to the implications of *water vapour* in the air. Section 2.7 recalls the basic thermodynamics of phase changes and introduces several measures of atmospheric water vapour content. These ideas are exploited in Section 2.8, in which some effects of the release of latent heat are investigated in a calculation of the *saturated adiabatic lapse rate*, which gives information on the stability of a moist atmosphere. The *tephigram*, a graphical method of representing the vertical structure of temperature and moisture and calculating useful physical results, is introduced in Section 2.9. Finally, some of the basic physics of the formation of cloud droplets by condensation of water vapour is considered in Section 2.10.

2.1 The ideal gas law

To a good approximation the atmosphere behaves as an ideal (or perfect) gas, with each mole of gas obeying the law

$$pV_{\mathrm{m}} = R^* T, \qquad (2.1)$$

where p is the pressure, V_m is the volume of one mole, R^* is the universal gas constant and T is the absolute temperature. We can obtain the corresponding law for *unit mass* of air by noting that, if the mass of one mole is M_m then the density $\rho = M_m/V_m$. So, from equation (2.1),

$$p = \frac{R^* T}{V_m} = \frac{R^*}{M_m} T\rho$$

and hence

$$p = RT\rho, \tag{2.2}$$

where $R \equiv R^*/M_m$ is the gas constant *per unit mass*. The value of R depends on the precise composition of the sample of air under consideration†.

2.2 Atmospheric composition

Consider a small sample of air of volume V, temperature T and pressure p, containing a mixture of gases G_i $(i = 1, 2, \dots)$. If there are n_i molecules of gas G_i in the sample, then the total number of molecules in the sample is

$$n = \sum n_i, \tag{2.3}$$

where the sum is taken over all the gases in the mixture, and the total mass of the sample is

$$m = \sum m_i n_i, \tag{2.4}$$

where m_i is the molecular mass of gas G_i.

We define the *mass mixing ratio* μ_i of gas G_i as the total mass of the molecules of gas G_i in the sample, divided by the total mass of the complete sample‡. Thus

$$\mu_i = \frac{m_i n_i}{m}. \tag{2.5}$$

We now introduce the ideal gas law in the form

$$pV = nkT, \tag{2.6}$$

where k is Boltzmann's constant. (The connection with the molar

† We follow the standard notation in atmospheric physics by using R for the gas constant per unit mass: this contrasts with other branches of physics, in which R usually denotes the molar gas constant. We use R^* for the latter.

‡ When the gas under consideration is water vapour, it may be more convenient, conceptually, to divide by the total mass *minus* the mass of water vapour, i.e., by the mass of dry air in the sample. However, in most cases when the mass mixing ratio is used it is a small number (e.g., <0.03 for water vapour and less still for other trace gases such as carbon dioxide and ozone), so the difference between the two definitions is usually insignificant.

form, equation (2.1), can be seen by noting that, for one mole, $n = N_A$, where N_A is Avogadro's number, and recalling that $R^* = N_A k$.) The *partial pressure* p_i of gas G_i is the pressure that would be exerted by the molecules of G_i from the sample if they alone were to occupy volume V at temperature T; from equation (2.6)

$$p_i = n_i \frac{kT}{V}. \tag{2.7}$$

Similarly, the *partial volume* V_i of gas G_i is the volume that would be occupied by the molecules of gas G_i from the sample if they, alone, were to be held at temperature T and pressure p; again from equation (2.6)

$$V_i = n_i \frac{kT}{p}. \tag{2.8}$$

(Note that Dalton's laws of partial pressures, $p = \sum p_i$, and partial volumes, $V = \sum V_i$, follow immediately from these definitions and equation (2.3).) From equations (2.5)–(2.7) we can relate the mass mixing ratio to the partial pressure as follows:

$$\mu_i = \frac{nm_i p_i}{mp} = \frac{m_i}{\overline{m}} \frac{p_i}{p}, \tag{2.9}$$

where

$$\overline{m} = m/n \tag{2.10}$$

is the mean molecular mass for the sample. We also define the *volume mixing ratio* v_i (also known as the *mole fraction*) by

$$v_i = \frac{V_i}{V};$$

by equations (2.6)–(2.8) we have

$$v_i = \frac{n_i}{n} = \frac{p_i}{p}. \tag{2.11}$$

Note that the two mixing ratios are related by

$$\mu_i = \frac{m_i}{\overline{m}} v_i.$$

Another measure of the concentration of an atmospheric gas is the *number density* (the number of molecules of the gas per unit volume), n_i/V. If we wish to follow the motion of the sample of air, the number density may change either through changes of the volume V of the sample or through changes of n_i resulting from chemical reactions. For many purposes the mass and volume mixing ratios are more convenient measures of concentration when the transport of chemicals is being studied, since they are affected not by volume changes but only by chemical production or loss. For other purposes, partial pressure is sometimes used to quantify chemical concentrations.

Table 2.1. *Some gases in the atmosphere. The unit ppmv (parts per million by volume) is used here for CO_2 and O_3; this is a standard unit of volume mixing ratio for minor species. The volume mixing ratios are fairly uniform throughout the lower and middle atmosphere for well-mixed gases that are mostly chemically inert, namely N_2, O_2, CO_2 and Ar. However, the volume mixing ratio for CO_2 is increasing by about 14 ppmv per decade: the value quoted is that for the late 1990s. The unit of molar mass is g mol^{-1} or equivalently kg kmol^{-1}.*

Gas	*Volume mixing ratio*	*Molar mass*	*Distribution*
Nitrogen, N_2	0.78	28.02	Well-mixed
Oxygen, O_2	0.21	32.00	Well-mixed
Carbon dioxide, CO_2	360 ppmv	44.01	Well-mixed
Water vapour, H_2O	$\lesssim 0.03$	18.02	Maximum in troposphere
Ozone, O_3	$\lesssim 10$ ppmv	48.00	Maximum in stratosphere
Argon, Ar	0.0093	39.95	Well-mixed

The Earth's atmosphere is composed mainly of nitrogen and oxygen, with a much smaller amount of carbon dioxide and still less of certain trace gases such as ozone. See Table 2.1 for a list of some of the more important species. We shall see in Chapters 3 and 6 that some gases such as carbon dioxide and ozone are of crucial importance in determining the structure of the atmosphere, despite the fact that they are present only in small amounts.

From equations (2.4), (2.10) and (2.11) the mean molecular mass of an air sample is

$$\overline{m} = \frac{m}{n} = \sum m_i \frac{n_i}{n} = \sum m_i \nu_i.$$

Similarly, the mean *molar* mass $\overline{M}$ is the mean of the molar masses M_i of the constituent gases G_i, weighted by the volume mixing ratios:

$$\overline{M} = \sum M_i \nu_i.$$

Using Table 2.1 it can be verified that the mean molar mass of dry air is about 28.97.

2.3 Hydrostatic balance

For an atmosphere at rest, in static equilibrium, the net forces acting on any small portion of air must balance. Consider for example a small cylinder of air, of height Δz and horizontal cross-sectional

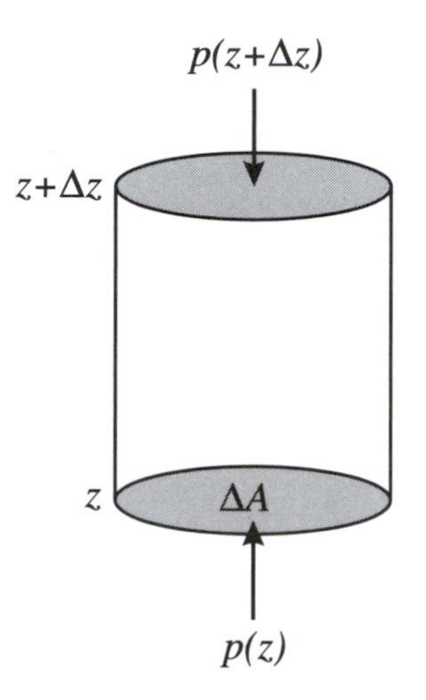

Figure 2.1 The vertical pressure forces acting on a small cylinder of air.

area ΔA. This is subject to a gravitational force $g\,\Delta m$ downwards, where its mass $\Delta m = \rho\,\Delta A\,\Delta z$ and g is the gravitational acceleration (assumed constant throughout this book); see Figure 2.1. This force must be balanced by the difference between the upward pressure force $p(z)\,\Delta A$ on the bottom of the cylinder and the downward pressure force $p(z+\Delta z)\,\Delta A$ on the top. We therefore have

$$g\rho\,\Delta A\,\Delta z = p(z)\,\Delta A - p(z+\Delta z)\,\Delta A;$$

by cancelling out ΔA and using the Taylor expansion

$$p(z+\Delta z) \approx p(z) + \frac{dp}{dz}\,\Delta z,$$

we get the equation for *hydrostatic balance*,

$$\frac{dp}{dz} = -g\rho. \tag{2.12}$$

(In Chapter 4 we extend this to the case in which the portion of air is accelerating and therefore no longer in static equilibrium.)

We can derive some basic properties of the atmosphere, given that it is an ideal gas and assuming that it is in hydrostatic balance. First eliminating the density ρ from equations (2.2) and (2.12), we obtain a useful alternative form of the hydrostatic balance equation,

$$\frac{dp}{dz} = -\frac{gp}{RT}. \tag{2.13}$$

If the temperature is a known function of height, $T(z)$, we can in principle find the pressure and density as functions of height also. For equation (2.13) can be rewritten

$$\frac{d}{dz}(\ln p) = -\frac{g}{RT}$$

and this may be integrated in z from the ground (say $z = 0$)

upwards, given the pressure at the ground (say p_0):

$$\ln p - \ln p_0 = -\frac{g}{R}\int_0^z \frac{dz'}{T(z')}$$

or, taking exponentials,

$$p = p_0 \exp\left(-\frac{g}{R}\int_0^z \frac{dz'}{T(z')}\right). \tag{2.14}$$

The simplest case is that of an isothermal temperature profile, i.e., $T = T_0 =$ constant, when the pressure decays exponentially with height:

$$p = p_0 \exp\left(-\frac{gz}{RT_0}\right) = p_0 e^{-z/H}, \tag{2.15}$$

where $H = RT_0/g$ is the *pressure scale height*, the height over which the pressure falls by a factor of e. In this isothermal case the density also falls exponentially with height in the same way: $\rho = \rho_0 \exp(-z/H)$, ρ_0 being the density at the ground. For an isothermal atmosphere with $T_0 = 260\,\mathrm{K}$, H is about 7.6 km.

The *lapse rate* Γ denotes the rate of decrease of temperature with height:

$$\Gamma(z) = -\frac{dT}{dz};$$

in general the temperature decreases with height ($\Gamma > 0$) in the troposphere and increases with height ($\Gamma < 0$) in the stratosphere; see Figure 1.3. A layer in which the temperature increases with height ($\Gamma < 0$) is called an *inversion layer*. If Γ is constant in the region between the ground and some height z_1, say, then the temperature in that region decreases linearly with height ($T(z) = T_0 - \Gamma z$, where T_0 is the ground temperature) and the integral in equation (2.14) can again be evaluated explicitly; see Problem 2.3.

Another useful deduction from the hydrostatic equation in the form (2.13) is the 'thickness', or depth, of the layer between two given surfaces of constant pressure. Suppose that the height of the pressure surface $p = p_1$ is z_1 and the height of the pressure surface $p = p_2$ is z_2. Then, if $p_1 > p_2$, we must have $z_1 < z_2$, since pressure decreases with height when hydrostatic balance applies. From equation (2.13), $g\,dz = -RT\,d(\ln p)$; integration gives

$$z_2 - z_1 = -\frac{R}{g}\int_{p_1}^{p_2} T\,d(\ln p).$$

The integral can in principle be evaluated if the temperature T is known as a function of pressure p: this may be provided for example by a weather balloon or a satellite-borne instrument. In particular, if T is constant,

$$z_2 - z_1 = \frac{RT}{g}\ln\left(\frac{p_1}{p_2}\right);$$

if T is not constant, we can still write

$$z_2 - z_1 = \frac{R\overline{T}}{g} \ln\left(\frac{p_1}{p_2}\right),$$

provided that we define $\overline{T}$ as a suitably weighted mean temperature within the layer:

$$\overline{T} = \frac{\int_{p_2}^{p_1} T\, d(\ln p)}{\int_{p_2}^{p_1} d(\ln p)}.$$

Thus the thickness of the layer between two pressure surfaces is proportional to the mean temperature of that layer.

2.4 Entropy and potential temperature

The First Law of Thermodynamics, applied to a small change to a closed system, such as a mass of air contained in a cylinder with a movable piston at one end (see Figure 2.2) can be written

$$\delta U = \delta Q + \delta W, \tag{2.16}$$

where δU is the increase of internal energy of the system in the process, δQ is the heat supplied to the system and δW is the work done on the system. In terms of functions of state, equation (2.16) can be written

$$\delta U = T\,\delta S - p\delta V, \tag{2.17}$$

where S is the *entropy* of the system. An alternative form of equation (2.17) is

$$\delta H = T\,\delta S + V\,\delta p, \tag{2.18}$$

where $H = U + pV$ is the *enthalpy*. Since equations (2.17) and (2.18) involve functions of state, they apply both for reversible and for irreversible changes. However, we shall mostly restrict our attention to reversible changes, so that the equations

$$\delta Q = T\,\delta S, \tag{2.19}$$
$$\delta W = -p\,\delta V$$

also hold.

For *unit mass* of ideal gas, for which $V = 1/\rho$, it can be shown that

$$U = c_v T, \tag{2.20}$$

where c_v is the specific heat capacity at constant volume and is independent of T. Therefore the ideal gas law, equation (2.2), implies that, for unit mass,

$$H = c_v T + RT = c_p T, \tag{2.21}$$

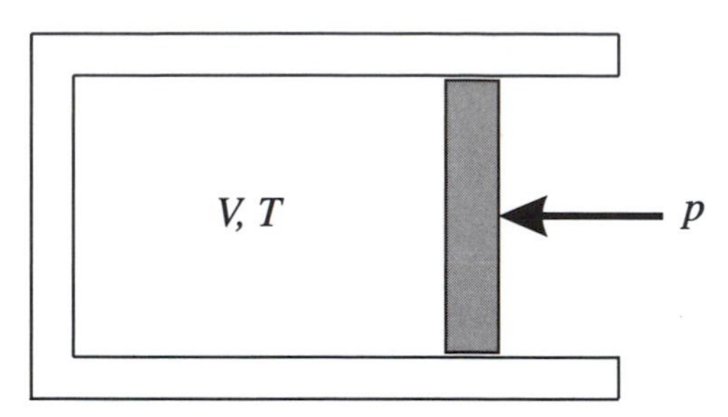

Figure 2.2 A cylinder of air of volume V, at pressure p and temperature T, closed by a movable piston (shaded).

where $c_p = c_v + R$ is the specific heat capacity at constant pressure. On substituting the expression (2.21) and $V = 1/\rho = RT/p$ into equation (2.18), we get

$$T\,\delta S = c_p\,\delta T - \frac{RT}{p}\,\delta p. \tag{2.22}$$

Division by T gives

$$\delta S = c_p\,\frac{\delta T}{T} - R\,\frac{\delta p}{p} = c_p\,\delta(\ln T) - R\,\delta(\ln p), \tag{2.23}$$

and integration gives the entropy per unit mass

$$S = c_p \ln T - R \ln p + \text{constant} = c_p \ln\left(T p^{-\kappa}\right) + S_0, \tag{2.24}$$

where $\kappa = R/c_p$, which is approximately $\frac{2}{7}$ for a diatomic gas, and S_0 is a constant.

An *adiabatic* process is one in which heat is neither gained nor lost, so that $\delta Q = 0$. Imagine a cylinder of air, originally at temperature T and pressure p, that is compressed adiabatically until its pressure equals p_0. We can find its resulting temperature, θ say, using equation (2.23); for, if $\delta Q = 0$ then $\delta S = 0$ also, from equation (2.19), so

$$c_p\,\delta(\ln T) = R\,\delta(\ln p).$$

Integrating and using the end conditions $T = \theta$ and $p = p_0$ then gives

$$c_p \ln\left(\frac{\theta}{T}\right) = R \ln\left(\frac{p_0}{p}\right),$$

and hence, using $\kappa = R/c_p$ again,

$$\theta = T\left(\frac{p_0}{p}\right)^{\kappa}. \tag{2.25}$$

The quantity θ is called the *potential temperature* of a mass of air at temperature T and pressure p. The value of p_0 is usually taken to be 1000 hPa. Using equation (2.24) it follows that the potential temperature is related to the specific entropy S by

$$S = c_p \ln \theta + S_1,$$

where S_1 is another constant. By definition, the potential temperature of a mass of air is constant when the mass is subject to an adiabatic change; conversely, the potential temperature will change when the mass is subject to a non-adiabatic (or *diabatic*) change. As we shall see, the potential temperature is often a very useful concept in atmospheric thermodynamics and dynamics.

2.5 Parcel concepts

We have just discussed adiabatic processes for a mass of air contained in a cylinder. To apply similar concepts to the atmosphere, we introduce the idea of an *air parcel* – a small mass of air that is imagined to be 'marked' in some way, so that its passage through the surrounding air ('the environment') can in principle be traced. The parcel is influenced by the environment, but we assume that it does not itself change the environment. The pressure within the parcel is taken to equal that of the surrounding environment, but its temperature, density and composition may differ from those of the environment. The parcel concept is useful, but should not be taken too literally; for example, a real mass of air will rapidly mix with its surroundings and will also inevitably influence the surrounding air.

One simple way to think of an air parcel is to imagine it to be enclosed in a thin balloon of negligible surface tension and heat capacity. We may also take the balloon to have negligible thermal conductivity, in which case the parcel moves adiabatically if there are no sources or sinks of heat within it†. In the adiabatic case, we can extend the definition of the potential temperature from a cylinder to a parcel of air; it is the final temperature θ of a parcel that is imagined to be brought adiabatically from pressure p and temperature T to pressure p_0.

For an adiabatically rising parcel, the potential temperature and entropy are constant as its height changes, so we can write

$$\left(\frac{d\theta}{dz}\right)_{\text{parcel}} = 0, \qquad \left(\frac{dS}{dz}\right)_{\text{parcel}} = 0.$$

From equation (2.23) we therefore have the following relation between the vertical derivatives of temperature and pressure, following the parcel:

$$0 = \frac{c_p}{T}\left(\frac{dT}{dz}\right)_{\text{parcel}} - \frac{R}{p}\left(\frac{dp}{dz}\right)_{\text{parcel}},$$

† Such heat sources could be due, for example, to latent heating or cooling; see Section 2.7.

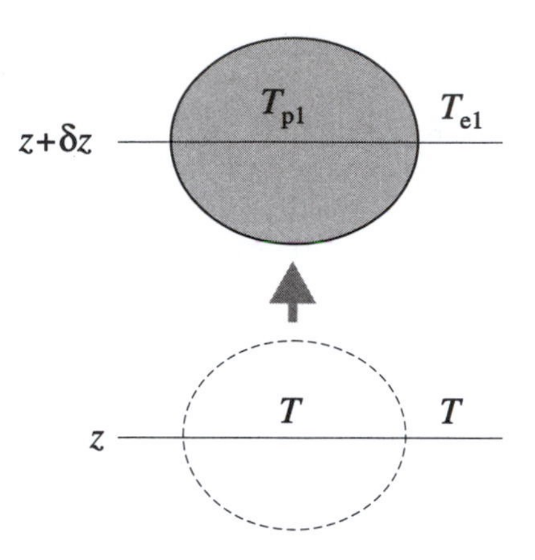

Figure 2.3 A parcel (shown shaded) displaced a height δz from its equilibrium position at height z (shown dashed).

so that

$$-\left(\frac{dT}{dz}\right)_{\text{parcel}} = -\frac{RT}{c_p p}\left(\frac{dp}{dz}\right)_{\text{parcel}} = \frac{g}{c_p} \equiv \Gamma_{\text{a}}, \tag{2.26}$$

say, where equations (2.12) and (2.1) have been used. The quantity Γ_{a} is the rate of decrease of temperature with height, following the adiabatic parcel as it rises. It is called the *adiabatic lapse rate*; when applied to a mass of dry air, it is called the *dry adiabatic lapse rate* (DALR) and is approximately 9.8 K km^{-1}.

An alternative derivation of the expression (2.26) for the DALR is to note that, for unit mass,

$$\delta Q = T\,\delta S = c_p\,\delta T - \frac{RT}{p}\,\delta p = c_p\,\delta T - \frac{\delta p}{\rho} = c_p\,\delta T + g\,\delta z, \tag{2.27}$$

from equations (2.19), (2.22), the ideal gas law (2.2) and the hydrostatic equation (2.12). For adiabatic motion of the parcel $\delta Q = 0$ and so, letting $\delta z \to 0$,

$$-\frac{dT}{dz} = \frac{g}{c_p} = \Gamma_{\text{a}},$$

as before.

The actual lapse rate $-dT/dz$ in the atmosphere will generally differ from the DALR. To investigate the implications of this, consider a parcel that is originally at equilibrium at height z, with temperature T, pressure p and density ρ, all equal to the values for the surroundings. Now suppose that an instantaneous upward force is applied to the parcel, so that it rises adiabatically through a small height δz, without influencing its surroundings; see Figure 2.3.

At the displaced position $z_1 = z + \delta z$ the parcel temperature has increased to T_{p1}, say, according to the adiabatic lapse rate:

$$T_{\text{p1}} = T + \left(\frac{dT}{dz}\right)_{\text{parcel}} \delta z = T - \Gamma_{\text{a}}\,\delta z. \tag{2.28}$$

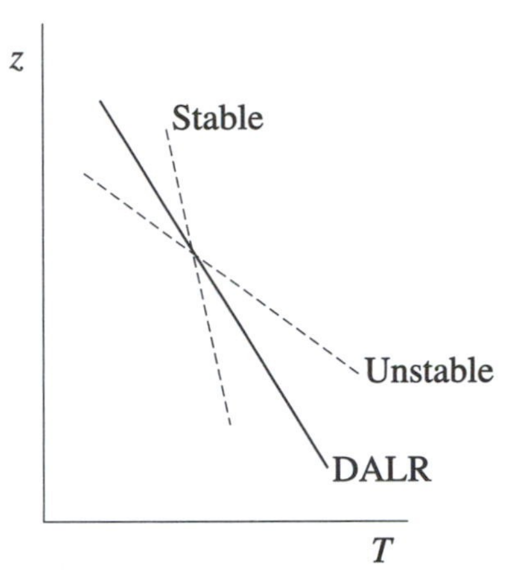

Figure 2.4 Schematic height profiles of temperature, showing the dry adiabatic lapse rate (DALR) and examples of stable and unstable environmental lapse rates.

On the other hand, the environment temperature at height z_1 is

$$T_{\mathrm{e}1} = T + \left(\frac{dT}{dz}\right)_{\mathrm{env}} \delta z = T - \Gamma\, \delta z. \tag{2.29}$$

If $\Gamma \neq \Gamma_{\mathrm{a}}$ there is therefore a temperature difference between the displaced parcel and its surroundings.

However, since the pressures are the same inside and outside the parcel at height z_1, these pressures are both equal to

$$p_1 = p + \left(\frac{dp}{dz}\right)_{\mathrm{env}} \delta z.$$

By the ideal gas law, equation (2.2), the densities inside and outside the parcel are

$$\rho_{\mathrm{p}1} = \frac{p_1}{RT_{\mathrm{p}1}}, \qquad \rho_{\mathrm{e}1} = \frac{p_1}{RT_{\mathrm{e}1}},$$

respectively. The volume of the parcel at height z_1 equals the volume of air displaced there; therefore, if $\rho_{\mathrm{p}1} > \rho_{\mathrm{e}1}$, the mass of the parcel at z_1 is greater than the mass of air displaced, so the parcel is 'heavier' than its surroundings. This holds provided that the temperature of the parcel is less than that of its surroundings ($T_{\mathrm{p}1} < T_{\mathrm{e}1}$), which in turn is true if $\Gamma < \Gamma_{\mathrm{a}}$, from equations (2.28) and (2.29), i.e., provided that the environment temperature falls less rapidly with height than the adiabatic lapse rate; see Figure 2.4. In this case the displaced parcel, being denser than its surroundings, will tend to fall back towards its equilibrium level; we say that the atmosphere is *statically stable* (or 'stable', for short) near height z.

On the other hand, if $\Gamma > \Gamma_{\mathrm{a}}$, so that the environment temperature falls more rapidly with height than the adiabatic lapse rate, a parcel displaced adiabatically upwards finds itself 'lighter' than its surroundings and hence continues to rise: the atmosphere is then said to be *statically unstable* (or 'unstable') near z; see Figure 2.4 again. If $\Gamma = \Gamma_{\mathrm{a}}$ we have neutral stability.

As a further step, we can investigate the buoyancy force on the parcel and its acceleration. The upward buoyancy force on the parcel is g times the difference between the mass of air displaced and the mass of the parcel, i.e.

$$gV_1(\rho_{\mathrm{e}1} - \rho_{\mathrm{p}1})$$

at height z_1, where V_1 is the volume of the parcel there. By Newton's Second Law, this force can be equated to the mass of the parcel times its acceleration, i.e.,

$$\rho_{\mathrm{p}1} V_1 \frac{d^2(\delta z)}{dt^2}.$$

We therefore have

$$\begin{aligned}\frac{d^2(\delta z)}{dt^2} &= g\left(\frac{\rho_{\mathrm{e}1}}{\rho_{\mathrm{p}1}} - 1\right) = g\left(\frac{T_{\mathrm{p}1}}{T_{\mathrm{e}1}} - 1\right) \\ &= g\left(\frac{T - \Gamma_{\mathrm{a}}\,\delta z}{T - \Gamma\,\delta z} - 1\right) = g\left(\frac{\Gamma - \Gamma_{\mathrm{a}}}{T - \Gamma\,\delta z}\right)\delta z \\ &= -\frac{g}{T}(\Gamma_{\mathrm{a}} - \Gamma)\,\delta z\end{aligned}$$

to leading order in the small quantity δz. (The ideal gas equation (2.2) has been used to go from the second to the third expression here.) We therefore have an equation of the form

$$\frac{d^2(\delta z)}{dt^2} + N^2(\delta z) = 0, \tag{2.30}$$

where

$$N^2 = \frac{g}{T}(\Gamma_{\mathrm{a}} - \Gamma) = \frac{g}{T}\left(\frac{dT}{dz} + \frac{g}{c_p}\right), \tag{2.31}$$

and the temperature T in equation (2.31) is that of the environment.

In the case of a statically stable region of the atmosphere, in which $\Gamma_{\mathrm{a}} > \Gamma$, equation (2.31) indicates that $N^2 > 0$ and equation (2.30) represents simple harmonic motion, with sinusoidal solutions: the parcel then oscillates up and down at an angular frequency N. N is called the *buoyancy frequency* or the *Brunt–Väisälä frequency*; for the lower atmosphere the corresponding period $2\pi/N$ is a few minutes (see Problem 2.6). For a statically unstable region of the atmosphere, in which $\Gamma_{\mathrm{a}} < \Gamma$, $N^2 < 0$ so that N is imaginary, which leads to exponential solutions of equation (2.30), one of which corresponds to the displaced parcel continuing to move at an increasing speed.

The quantity N^2 is a useful measure of atmospheric stratification. It can be related to the potential temperature of the environment as follows. Taking logarithms of equation (2.25) and differentiating, we obtain

$$\frac{1}{\theta}\frac{d\theta}{dz} = \frac{1}{T}\frac{dT}{dz} - \frac{\kappa}{p}\frac{dp}{dz} = \frac{1}{T}\frac{dT}{dz} + \frac{\kappa\rho g}{p}$$
$$= \frac{1}{T}\frac{dT}{dz} + \frac{g}{c_p T}, \qquad (2.32)$$

using the ideal gas law (2.2) and the relation $c_p = R/\kappa$. On combining equations (2.31) and (2.32) we obtain the equation

$$N^2 = \frac{g}{\theta}\frac{d\theta}{dz}.$$

Thus a region of the atmosphere is statically stable if θ increases with height and is statically unstable if θ decreases with height†.

2.6 The available potential energy

It is frequently useful to consider the energy content of the atmosphere, so some basic concepts are presented here. Consider first the potential energy E_{P} of a vertical column of atmosphere, of unit horizontal cross-section, stretching from the ground (where $z = z_0$ and $p = p_0$, say) to a great height (where $z = \infty$ and $p = 0$). Since the potential energy of a small slice, of thickness Δz, of this column is $(\rho\,\Delta z)gz$ the potential energy of the whole column is

$$E_{\mathrm{P}} = \int_{z_0}^{\infty} \rho g z\, dz.$$

However, $\rho g = -dp/dz$ from the hydrostatic equation (2.12), so we can write

$$E_{\mathrm{P}} = -\int_{z_0}^{\infty} z\frac{dp}{dz}\,dz = \int_{p=0}^{p=p_0} z\, dp,$$

where the integration variable has been changed from z to p and the upper and lower limits of integration have been reversed, with a corresponding change of sign. We can integrate the last expression in this equation by parts to get

$$E_{\mathrm{P}} = \left[zp\right]_{p=0}^{p=p_0} + \int_{z_0}^{\infty} p\, dz.$$

Now assuming‡ that $zp \to 0$ as $z \to \infty$ and $p \to 0$, we get

$$E_{\mathrm{P}} = \int_{z_0}^{\infty} p\, dz + E_0, \qquad \text{where} \qquad E_0 = z_0 p_0. \qquad (2.33)$$

† Note that, for a compressible atmosphere, the static stability depends on the vertical increase of the potential temperature, *not* on the vertical decrease of the density, as might have been expected intuitively. The latter is applicable only to an incompressible fluid. See Problem 2.7.

‡ This holds in all reasonable cases; for example, p decays exponentially with z in an isothermal atmosphere, as shown in equation (2.15).

The boundary term $E_0 = 0$ if the surface elevation $z_0 = 0$, but in general we must allow for topography of varying height over the Earth's surface, with $z_0 \neq 0$.

We now consider the internal energy of the air in the column. As noted in equation (2.20) the internal energy per unit mass is $c_v T$, so the integral of this over the mass of the column is

$$E_{\mathrm{I}} = \int_{z_0}^{\infty} \rho c_v T \, dz = \frac{c_v}{R} \int_{z_0}^{\infty} p \, dz, \tag{2.34}$$

where the ideal gas law (2.2) has been used. Note that E_{I} is proportional to the integral in equation (2.33). We define the *total potential energy* E_{T} of the column as the sum of the potential and internal energies so, from equations (2.33) and (2.34),

$$E_{\mathrm{T}} = E_{\mathrm{P}} + E_{\mathrm{I}} = \frac{c_v + R}{R} \int_{z_0}^{\infty} p \, dz + E_0 = \frac{1}{\kappa} \int_{z_0}^{\infty} p \, dz + E_0,$$

since $c_v + R = c_p = R/\kappa$. Note that the integral here is the total enthalpy of the column, since the enthalpy per unit mass is $c_p T$, from equation (2.21), so that the enthalpy per unit volume is $c_p \rho T = c_p p/R$, using the ideal gas law, equation (2.2).

We can extend these ideas to the atmosphere as a whole by integrating the column values over the Earth's surface area; for example the global total energy is

$$\langle E_{\mathrm{T}} \rangle = \iint_{\text{surface}} E_{\mathrm{T}} \, dx \, dy,$$

where x and y are horizontal coordinates.

Given any atmospheric state, we need to ask whether a suitable redistribution of mass can lead to a state of lower total potential energy, thus releasing energy that can be converted, say, to kinetic energy of motion. The answer in general is yes, but it turns out that the maximum amount of potential energy that can be released is in practice very much less than $\langle E_{\mathrm{T}} \rangle$. Lorenz showed how the state of minimum total potential energy (or *reference state*) can be calculated, assuming an *adiabatic* redistribution of the atmospheric mass.

A crucial part of Lorenz's calculation was to note that, given an adiabatic redistribution of mass, the mass of air above any given surface of constant potential temperature θ (an *isentropic surface* or *isentrope*) will not change. This is because, in an adiabatic process, each small mass element of air must retain its potential temperature and must therefore remain on a given isentropic surface. Consider then the mass M_1 above the isentrope $\theta = \theta_1$, say, in the actual atmospheric state; this is

$$M_1 = \iint_{\text{surface}} dx \, dy \int_{z_1}^{\infty} \rho \, dz,$$

where $z_1(x, y)$ is the height of the isentrope. From the hydrostatic equation (2.12) this equals

$$M_1 = \frac{1}{g}\iint_{\text{surface}} dx\,dy \int_0^{p_1} dp = \frac{1}{g}\iint_{\text{surface}} p_1(x, y)\,dx\,dy = \frac{1}{g}\langle p_1\rangle,$$

where $p_1(x, y)$ is the pressure on the isentrope†. In particular the total mass of the atmosphere is

$$M_s = \frac{1}{g}\iint_{\text{surface}} p_0(x, y)\,dx\,dy = \frac{1}{g}\langle p_0\rangle.$$

The same results must hold for the reference state so that, under an adiabatic redistribution of mass, the global integral $\langle p_1\rangle$ (or global mean) of the pressure on the θ_1 isentrope (and therefore on every isentrope) must remain unchanged.

Together with other arguments, this allows the reference state and also the difference in total potential energy between the actual and reference states to be calculated. This difference is called the *available potential energy* and represents the maximum amount of potential energy that might be released for conversion into, say, kinetic energy. The details of the calculation are generally quite complicated; however a simple special case is considered in Problem 2.8.

2.7 Moisture in the atmosphere

It was mentioned in Section 2.2 that water vapour is a minor constituent of the atmosphere. However, despite its low and variable volume mixing ratio ($\lesssim 0.03$), water vapour plays a crucial role in weather and climate processes. It is responsible for the precipitation (rain and snow) that forms such an important part of our weather; it has significant radiative effects (see Chapter 3); and, through polar stratospheric clouds, it is associated, in its ice phase, with important chemical reactions that lead to the depletion of polar ozone (see Chapter 6).

In the remainder of this chapter we consider the precipitation properties of water vapour; these result from the physical fact that, at terrestrial atmospheric temperatures and pressures, water undergoes *phase changes* between the vapour, liquid and solid states.

We recall some basic thermodynamics of phase changes, focusing first on the vapour–liquid phase transition at a plane interface. At

† This is the global integral of the result that, under hydrostatic balance, the pressure at a given level equals g times the mass per unit horizontal area of air above that level. See also Problem 2.1.

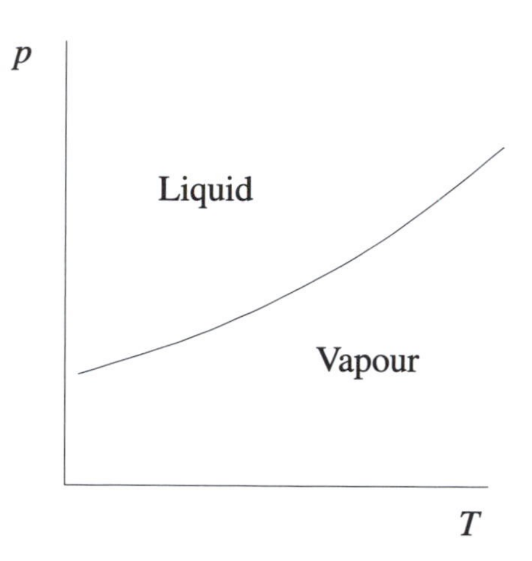

Figure 2.5 The phase transition between liquid and vapour, portrayed in a schematic temperature–pressure diagram. The curve indicates the temperature variation of the saturation vapour pressure.

this transition, the vapour is *saturated*†, that is, in equilibrium with the liquid. The rates of evaporation and condensation are equal, with as many molecules escaping from unit area of the liquid surface per unit time as return. The phase transition is conveniently portrayed in a p–T diagram (Figure 2.5), which indicates the temperature variation of the pressure at which the phase transition takes place (the *equilibrium vapour pressure* or the *saturation vapour pressure*, SVP). The slope of this curve is given by the *Clausius–Clapeyron equation*

$$\frac{dp}{dT} = \frac{\delta S}{\delta V} = \frac{L}{T\,\delta V}; \tag{2.35}$$

where δS is the entropy gained as unit mass of water changes from liquid to vapour (it is positive, since the vapour exhibits greater disorder than does the liquid state), δV is the increase of volume as unit mass changes from liquid to vapour (it is also positive) and L is the latent heat of vaporization per unit mass (also called the specific enthalpy of vaporization), given by $L = T\,\delta S$. In fact the specific volume V_v of the water vapour is much greater than that of the liquid, so $\delta V \approx V_v = 1/\rho_v = R_v T/p$, using the ideal gas law, equation (2.2), for the vapour; here ρ_v is the vapour density and R_v is the specific gas constant for the vapour. Hence the Clausius–Clapeyron equation can be written in the more convenient form

$$\frac{dp}{dT} = \frac{Lp}{R_v T^2}, \tag{2.36}$$

where p and T refer to the values at the phase transition.

These considerations apply to water vapour on its own, but can

† See Bohren (1987) for a clear exposition of the misleading nature of the adjective *saturated* in this context: in particular for its suggestion that air can 'only hold' a certain amount of water vapour.

also be applied to water vapour in the presence of 'air', provided that p is replaced by the partial pressure of water vapour, traditionally denoted by the symbol e. Using also the notation $e_s(T)$ to denote the saturation vapour pressure (at the phase transition), we can write the Clausius–Clapeyron equation (2.36) in the form

$$\frac{de_s}{dT} = \frac{Le_s}{R_v T^2}. \tag{2.37}$$

Note that if L is constant (a fairly good approximation at typical atmospheric temperatures), this can be integrated to give

$$e_s(T) = e_1 \exp\left(-\frac{L}{R_v T}\right), \tag{2.38}$$

where e_1 is a constant.

We can relate the partial pressure e of water vapour to its volume mixing ratio v and mass mixing ratio μ, say, using equations (2.11) and (2.9):

$$v = \frac{e}{p}, \qquad \mu = \epsilon\frac{e}{p}, \tag{2.39}$$

where p is the total air pressure, as usual. Here

$$\epsilon = \frac{m_v}{\overline{m}} \approx \frac{m_v}{m_d} = \frac{18.02}{28.97} = 0.622.$$

Here m_v is the molecular mass of water vapour, $\overline{m}$ is the mean molecular mass of moist air and m_d is the molecular mass of dry air (i.e., the constituents of the air other than water vapour). (Molar masses may also be used here.) Since the mixing ratio of the water vapour is so small ($\mu < 3 \times 10^{-2}$), $\overline{m} \approx m_d$. Another important measure of the moisture content of air is the *relative humidity*, defined by

$$\mathrm{RH} = \frac{e}{e_s(T)}$$

and usually expressed as a percentage.

Now consider a parcel of moist air, of unit mass, containing mass μ of water vapour, so that μ is the mass mixing ratio of water vapour according to our definition (2.5). So long as no condensation or evaporation takes place, this mass μ remains constant. The vapour pressure of the parcel satisfies

$$e = \frac{\mu p}{\epsilon} \tag{2.40}$$

by equation (2.39) and while this remains less than $e_s(T)$ saturation does not occur, since the water vapour in the parcel remains below the vapour pressure curve in Figure 2.5. Suppose that the parcel rises adiabatically from the surface (at pressure p_0 and temperature

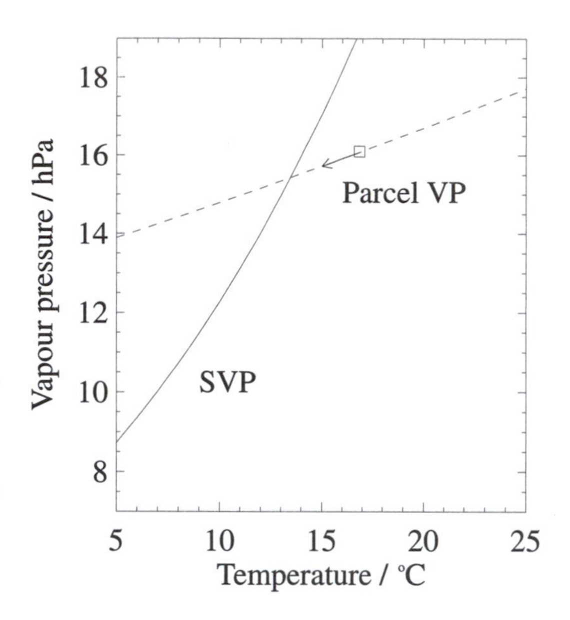

Figure 2.6 Saturation vapour pressure (SVP) curve (solid) and vapour pressure curve for a parcel of air containing water vapour of mass mixing ratio $= 10^{-2} = 10\,\mathrm{g\,kg^{-1}}$ (dashed). The parcel is taken to start at the point marked by a square, at which p_0 = 1000 hPa and T_0 = 290 K $\approx 17\,°\mathrm{C}$. As the parcel rises, its vapour pressure moves down the dashed curve in the direction of the arrow, intersecting the SVP curve at $T \approx$ 13.3 °C.

T_0); then the potential temperature of the parcel remains constant at $\theta = T_0$, while the temperature T of the parcel falls according to

$$T = T_0 \left(\frac{p}{p_0} \right)^{\kappa}; \qquad (2.41)$$

cf. equation (2.25). From equations (2.40) and (2.41) we can eliminate the pressure p of the parcel to find how the vapour pressure varies as a function of temperature T, following the motion of the parcel:

$$e_{\mathrm{parcel}}(T) = \frac{\mu p_0}{\epsilon} \left(\frac{T}{T_0} \right)^{1/\kappa}.$$

As illustrated in Figure 2.6, eventually the temperature of the rising parcel falls enough for e_{parcel} to equal e_{s} and saturation occurs. (However, it should be noted that in practice liquid need not form at saturation; the vapour may become supersaturated. It is usually necessary for small 'condensation nuclei' to be present before liquid drops appear; see Section 2.10.)

A useful related concept is the *saturation mixing ratio*, defined as

$$\mu_{\mathrm{s}}(T, p) = \frac{e_{\mathrm{s}}(T)\epsilon}{p} \qquad (2.42)$$

and conveniently measured in units of grammes per kilogramme ($\mathrm{g\,kg^{-1}}$). If, at temperature T and pressure p, the mixing ratio μ is less than $\mu_{\mathrm{s}}(T, p)$, then $e < e_{\mathrm{s}}$ from equations (2.40) and (2.42) and the air is unsaturated; it is saturated if $\mu = \mu_{\mathrm{s}}(T, p)$ and is supersaturated if $\mu > \mu_{\mathrm{s}}(T, p)$. A plot of μ_{s} as a function of temperature

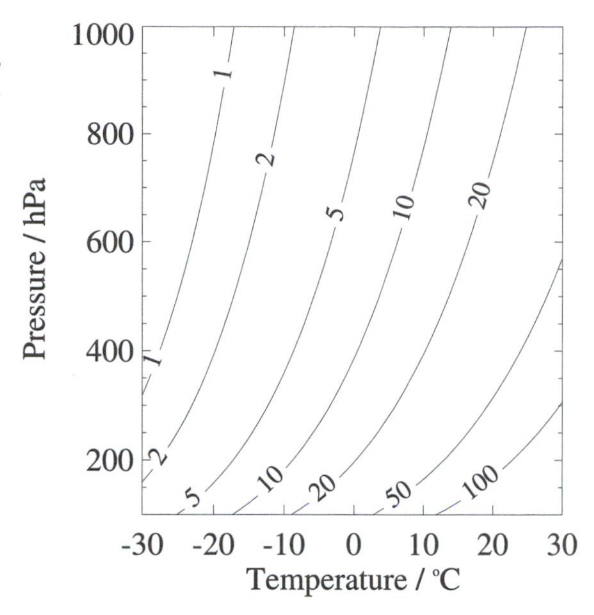

Figure 2.7 Contours of the saturation mixing ratio $\mu_s(T, p)$, in units of $g\,kg^{-1}$.

and pressure is given in Figure 2.7. The overall behaviour of μ_s is clearly consistent with the facts that it is inversely proportional to pressure and proportional to e_s, which increases with temperature (see Figure 2.6).

The *dew point* T_d of a sample of air is the temperature to which the air must be cooled *at constant pressure* (i.e., *not* following a rising parcel), retaining its water vapour content, for it to become saturated. Therefore, if the water vapour mixing ratio is μ, the dew point T_d satisfies the implicit equation

$$\mu_s(T_d, p) = \mu \, . \tag{2.43}$$

Equivalently, if the air sample initially has vapour pressure e, then $e_s(T_d) = e$.

We now briefly mention the ice phase. In addition to the vapour–water phase transition just considered, there are also ice–water and ice–vapour transitions, as shown in Figure 2.8. The three transition lines meet at the *triple point*, at the temperature $T_t = 273\,K$ and pressure $p_t = 6.1\,hPa$. The densities and, therefore, the specific volumes V of the three phases are different at the triple point: the specific volume of liquid water is $V_l = 1.00 \times 10^{-3}\,m^3\,kg^{-1}$, the specific volume of ice is $V_i = 1.09 V_l$ (unlike most substances, water expands on freezing) and the specific volume of water vapour is $V_v \approx 2 \times 10^5 V_l$.

The basic form (2.35) of the Clausius–Clapeyron equation applies to each of these transitions. Since $V_v \gg V_i$ and the latent heat of sublimation (ice–vapour) is approximately constant, an approximate form of the sublimation curve can be found as in equation (2.38);

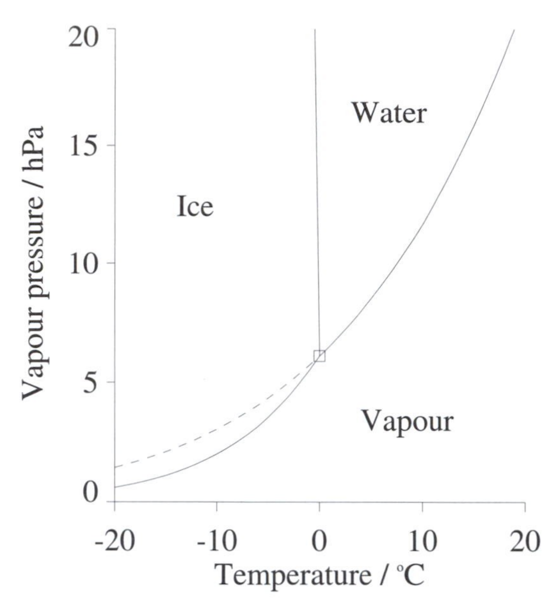

Figure 2.8 Schematic diagram showing the phase transitions between ice, liquid water and water vapour. The difference between the ice-vapour and vapour-water curves below 0 °C (where the vapour-water curve is shown dashed) has been exaggerated. Note that the ice-water curve is not quite vertical, but has a large negative slope. The triple point is indicated by the small square.

indeed, since the two latent heats are fairly similar in size, the two curves are quite close. However, the form of the ice–water transition curve is very different: here δV is small and negative, since V_l is slightly less than V_i, but the latent heat of fusion is positive and non-negligible. Hence the ice–water transition has a large negative slope, as indicated in Figure 2.8.

Many of the concepts developed above for condensation of water vapour at the vapour–liquid transition can be applied also to sublimation at the vapour–ice transition. These include the saturation vapour pressure, saturation mixing ratio and the *frost point*: the temperature to which moist air must be cooled at constant pressure for sublimation to occur.

2.8 The saturated adiabatic lapse rate

While the air in a rising parcel remains unsaturated, the derivation of the adiabatic lapse rate (see equation (2.26)) remains unchanged, apart from the use of the specific heat capacity c_p for the mixture of dry air and water vapour: this is always close to that for dry air alone (see Problem 2.5). However, once saturation takes place, the calculation of the lapse rate following the parcel must be changed significantly, because of the latent heat released.

We extend the second derivation of Section 2.5, beginning with equation (2.27), relating the heat input into the parcel (which is taken for convenience to be of unit mass) while it rises a distance

δz and its temperature increases by δT:

$$\delta Q = c_p \, \delta T + g \, \delta z, \tag{2.44}$$

where c_p is the value for the dry air–water vapour mixture. At saturation, the mixing ratio μ equals the saturation mixing ratio $\mu_s(T, p)$. If a mass $|\delta \mu_s| = -\delta \mu_s$ of water then condenses[†] during the rise through height δz an amount of latent heat

$$\delta Q = -L \, \delta \mu_s \tag{2.45}$$

is given to the gas. The liquid water is assumed to fall out of the parcel and take no further part in its heat balance: this is an *irreversible* process and it also implies that the air in the parcel undergoes a non-adiabatic change. However, the amount of heat removed from the parcel by the liquid water is small compared with that remaining in the parcel, so the process is referred to as *pseudo-adiabatic*.

On equating the two expressions for δQ in equations (2.44) and (2.45) we obtain

$$c_p \, \delta T + g \, \delta z + L \, \delta \mu_s = 0. \tag{2.46}$$

We now need to express $\delta \mu_s$ in terms of δT and δz. From equation (2.42) we have $\mu_s = \epsilon e_s / p$. Taking logarithms and differentiating gives

$$\frac{\delta \mu_s}{\mu_s} = \frac{\delta e_s}{e_s} - \frac{\delta p}{p}.$$

However, e_s depends only on T, so $\delta e_s = (de_s/dT)\,\delta T$; moreover, from the Clausius–Clapeyron equation (2.37),

$$\frac{1}{e_s}\frac{de_s}{dT} = \frac{L}{R_v T^2}.$$

From the hydrostatic equation in the form (2.13) we have $\delta p = -g\,\delta z/(RT)$, where p is the total pressure; by collecting these results we therefore get

$$\frac{\delta \mu_s}{\mu_s} = \frac{L \, \delta T}{R_v T^2} + \frac{g \, \delta z}{RT}. \tag{2.47}$$

By eliminating $\delta \mu_s$ from equations (2.46) and (2.47), we obtain

$$\left(c_p + \frac{L^2 \mu_s}{R_v T^2} \right) \delta T + g \left(1 + \frac{L \mu_s}{RT} \right) \delta z = 0.$$

Letting $\delta z \to 0$, we get the *saturated adiabatic lapse rate*[‡] (SALR)

[†] Note that $\delta \mu_s$, as is usual for a small change, is defined as an *increase* in μ_s; therefore $-\delta \mu_s$ is a *decrease* of μ_s.

[‡] Or, more correctly, the *pseudo-adiabatic lapse rate*.

Γ_s:

$$\Gamma_s = -\frac{dT}{dz} = \frac{g}{c_p}\frac{\left(1+\dfrac{L\mu_s}{RT}\right)}{\left(1+\dfrac{L^2\mu_s}{c_p R_v T^2}\right)}. \tag{2.48}$$

Note that the factor g/c_p on the right-hand side of equation (2.48) equals the DALR Γ_a.

For typical atmospheric values of T and μ_s it is found that $\Gamma_s \leq \Gamma_a$. Because of the latent heat given to the air by condensation of the water vapour, the temperature drops off less rapidly with height (by about 6–9 K km^{-1}) at the SALR than it does at the DALR (~ 9.8 K km^{-1}, as noted in Section 2.5); see Problem 2.9. Note that, through its dependences on T and $\mu_s(T,p)$, Γ_s depends on the temperature and pressure. Working in terms of the pressure of the parcel, rather than its height, we may show (again using $g\,\delta z = -RT\,\delta p/p$) that, following the ascending parcel,

$$\frac{dT}{dp} = \frac{\Gamma_s RT}{gp} = \Gamma_s'(T,p), \tag{2.49}$$

say. Curves in the T, p plane whose slopes at each point are given by equation (2.49) are called *saturated adiabatics*. Given the expression for Γ_s' and suitable starting values of T and p, they may readily be calculated numerically.

The arguments of Section 2.5 relating to the static stability of a region of the atmosphere can now be extended to allow for the effects of moisture. In particular, it follows that, if the actual lapse rate Γ is less than the SALR Γ_s, then the region is statically stable even if the air is saturated. However, if $\Gamma > \Gamma_s$, a saturated parcel will be unstable. Moreover, if $\Gamma_s < \Gamma < \Gamma_a$, a saturated parcel is unstable but an unsaturated one is not: this situation is called *conditional instability*.

Under a certain approximation, the saturated adiabatics can be calculated explicitly. First, note that, for reversible processes, $T\,\delta S = \delta Q$, so equation (2.23) can be written

$$c_p\,\delta(\ln T) - R\,\delta(\ln p) = \frac{\delta Q}{T} = -\frac{L\,\delta\mu_s}{T}. \tag{2.50}$$

Provided that the expression on the extreme right-hand side of equation (2.50) can be approximated by $-\delta(L\mu_s/T)$†, then

$$\delta\left(c_p \ln T - R\ln p + \frac{L\mu_s}{T}\right) = 0.$$

† This may be shown to hold if $L\mu_s/(c_pT) \ll 1$; since $L/(c_pT) \lesssim 10$ for typical lower-atmospheric temperatures, this approximation holds when $\mu_s \ll 100$ g kg^{-1}. From Figure 2.7 it can be seen that this is true over a wide range of temperatures and pressures. See also Problem 2.9.

On integrating, dividing by c_p, using $\kappa = R/c_p$ and taking exponentials, we get

$$\theta_e(T,p) \equiv T\left(\frac{p}{p_0}\right)^{-\kappa} \exp\left(\frac{L\mu_s}{c_p T}\right) = \text{constant}. \qquad (2.51)$$

The quantity θ_e is called the *equivalent potential temperature*. By comparison of equations (2.25) and (2.51) we see that

$$\theta_e(T,p) = \theta(T,p)\exp\left(\frac{L\mu_s(T,p)}{c_p T}\right).$$

Under the given approximation, we have therefore integrated equation (2.49) explicitly, so the curves of constant θ_e closely approximate the saturated adiabatics. It may be shown that, as we follow a saturated adiabatic $\theta_e = \theta_0$, say, to low pressure (and low temperature), it approaches the dry adiabatic $\theta = \theta_0$.

2.9 The tephigram

Meteorologists find it convenient to represent the vertical profiles of atmospheric temperature and moisture on *thermodynamic diagrams*. These profiles may be measured for example by an ascending instrumented balloon (a radiosonde). Thermodynamic diagrams are particularly useful for examining the effects of moisture, for which there are no simple formulae allowing easy analytical calculations.

One such diagram is the *tephigram*. This uses the temperature T and entropy per unit mass S as orthogonal coordinates†. The lines of constant S are equispaced in S but are labelled with the corresponding values of the potential temperature θ (recall that $S = c_p \ln\theta$). They are called *dry adiabatics*; the lines of constant T are called *isotherms*. Since

$$p = p_0\left(\frac{T}{\theta}\right)^{1/\kappa},$$

curves of constant pressure can also be plotted; in the range of T and θ relevant for the lower atmosphere, these are almost straight. The isotherms (parallel to the S axis) and the dry adiabatics (parallel to the T axis) are chosen to point at 45° above and below the horizontal, respectively, so that the curves of constant p (*isobars*) become roughly horizontal. With these curves plotted, the tephigram takes the form of Figure 2.9.

Suppose that an imaginary air parcel is moved reversibly around a closed circuit C in the tephigram, i.e. that it is forced through a cyclical process in which its temperature, entropy and, therefore,

† In older works, ϕ was used instead of S, hence the name T–ϕ diagram, or tephigram.

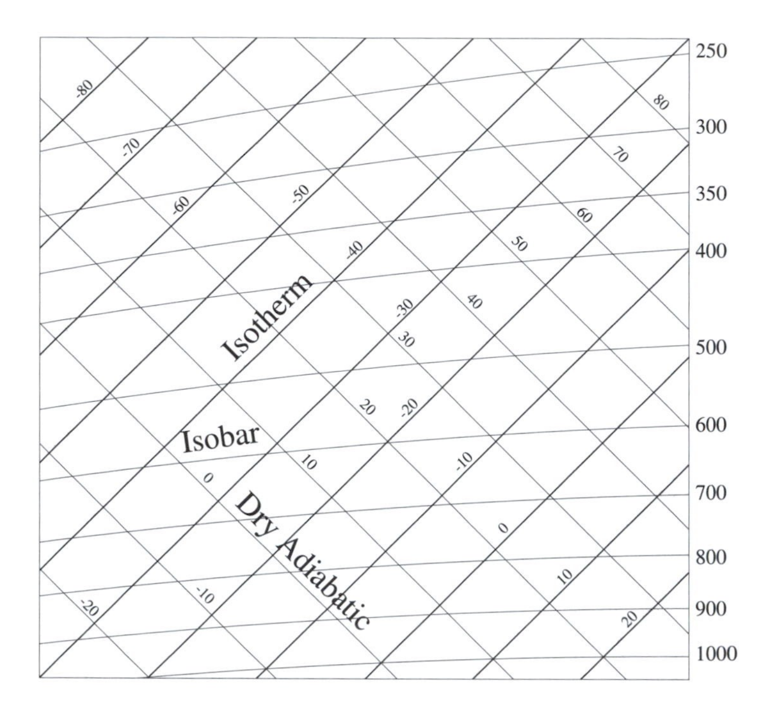

Figure 2.9 A partially completed tephigram, showing isotherms (°C), dry adiabatics (°C) and isobars (hPa).

pressure are varied, as shown in Figure 2.10. Since this process is cyclical, the total change of internal energy, δU say, is zero; but by the First Law of Thermodynamics, equation (2.17), $\delta U = T\,\delta S - p\,\delta V$, where V is the volume of the parcel. Hence the work done *on* the parcel, $-\oint_C p\,dV$ is given by $-\oint_C T\,dS$, which equals the area enclosed by the circuit (described clockwise) in the tephigram.

We can now add two other sets of curves, related to moisture. The first are the lines of constant saturation mixing ratio $\mu_s(T, p)$: these are almost straight and are drawn dashed. (Note that we can if we wish use T and p as independent variables for plotting points on the tephigram, instead of T and S.) They can be plotted using equation (2.42) if an accurate expression for $e_s(T)$ is known. Also plotted are the saturated adiabatics; these are noticeably curved. Each saturated adiabatic can be labelled by the temperature at which it cuts the $p = 1000\,\text{hPa}$ surface (the *wet-bulb potential temperature* θ_w) or the potential temperature which it approaches at low p (the equivalent potential temperature θ_e). Figure 2.11 shows the completed tephigram.

During the ascent of a radiosonde the temperature is measured at a series of pressure levels, which can be plotted on the tephigram, to give the *environment curve*. Straight-line segments are drawn between each point, rather than a smooth curve. This gives a representation of a vertical column of atmosphere – but note that this may be slightly misleading, since the balloon takes some time

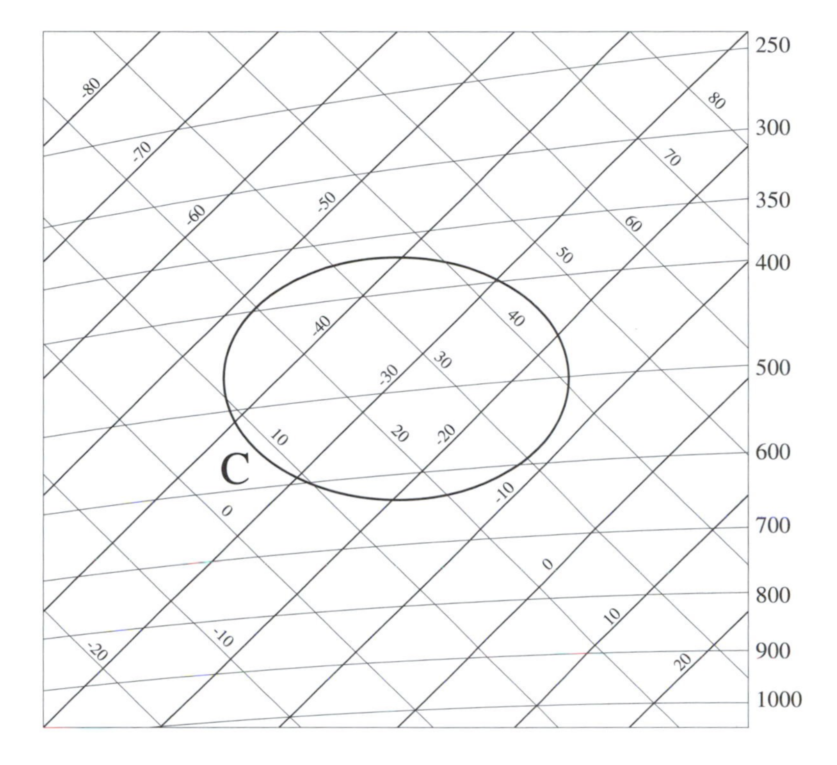

Figure 2.10 The same as figure 2.9, but with a closed path C marked.

to ascend and also blows some distance downwind as it does so. The dew point is also plotted at each pressure level, giving a separate curve. Equation (2.43) then allows us to find the mixing ratio μ at any pressure level on the environment curve, given the corresponding dew point: we just read off the value of μ_s at the same pressure on the dew point curve.

Many useful results can be obtained from the environment and dew-point curves, including inferences about the formation of clouds and the onset of instability. Examples are given in Problems 2.10–2.12 and students wishing to understand the tephigram should attempt some of these.

2.10 Cloud formation

In the previous subsections we have considered only the case of a flat interface between liquid and vapour. However, this is not directly relevant for the formation of cloud droplets, which are approximately spherical. We shall find that it is essential to allow for the effects of surface tension on cloud droplets. A further finding is that large supersaturation (a vapour pressure e significantly greater than the SVP e_s, or a relative humidity significantly greater than 100%) is required if cloud droplets are to form spontaneously from

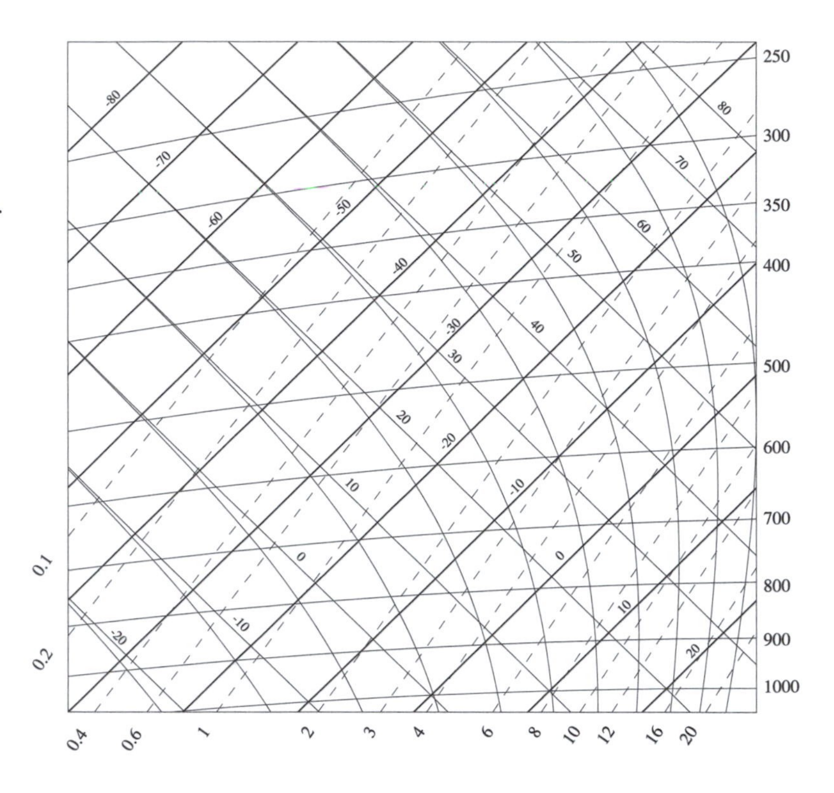

Figure 2.11 The completed tephigram. As Figure 2.9, but with lines of constant saturation mixing ratio (g kg^{-1}, dashed) and saturated adiabatics (curved) added.

water vapour: in fact small particles (*cloud-condensation nuclei*) are usually needed in the formation process.

For including the effects of surface tension, we introduce the *Gibbs free energy*

$$G = U - TS + pV$$

and note that, using equation (2.16),

$$\delta G = -S\,\delta T + V\,\delta p. \tag{2.52}$$

Consider a water droplet immersed in water vapour, at partial pressure e and temperature T (the remaining 'dry air' plays no role and can be ignored). Suppose that the liquid and vapour are not in equilibrium, so that $e \neq e_s(T)$, the SVP†. Suppose that the Gibbs free energy *per unit mass* of the vapour is $G_v(T, e)$ and that of the liquid is $G_l(T, e)$. Now let the partial pressure be varied slightly from e to $e + \delta e$, while the temperature is held constant. From equation (2.52) it follows that there are small changes to G_v and G_l, given by

$$\delta G_v = V_v\,\delta e, \qquad \delta G_l = V_l\,\delta e,$$

† It can be shown that the difference in pressure between the droplet and the surrounding vapour, due to surface tension, can be ignored here; see Salby (1996), Section 9.2.1.

where V_v and V_l are the specific volumes of the vapour and liquid, respectively. However, $V_v \gg V_l$, hence

$$\delta(G_v - G_l) = (V_v - V_l)\,\delta e \approx V_v\,\delta e.$$

The vapour satisfies the ideal gas law $V_v = R_v T/e$, so

$$\delta(G_v - G_l) = R_v T\,\frac{\delta e}{e} = R_v T\,\delta(\ln e).$$

Integrating at fixed T gives

$$G_v(T,e) - G_l(T,e) = R_v T \ln e + F(T),$$

where $F(T)$ is a function of integration. However, at equilibrium, on the vapour pressure curve where $e = e_s(T)$, the specific Gibbs free energies G_v and G_l are equal. (This result is used in the standard derivation of the Clausius–Clapeyron equation.) Using this condition to fix $F(T)$, we get

$$G_v(T,e) - G_l(T,e) = R_v T \ln\left(\frac{e}{e_s(T)}\right). \tag{2.53}$$

Now suppose that at some initial time we have a mass M_0 of water vapour, at partial pressure e and temperature T, with no droplet present. The total Gibbs free energy at this time is

$$\mathcal{G}_0 = G_v(T,e)M_0. \tag{2.54}$$

A droplet then starts to condense, at fixed temperature and pressure; suppose that at some later instant its radius is a, so that its surface area is $A = 4\pi a^2$ and its mass is $M_l = 4\pi a^3\rho_l/3$ (where ρ_l is the density of the liquid) and the mass of the surrounding vapour is M_v. The total Gibbs free energy of the system is now the sum of the Gibbs free energies of the liquid and vapour, plus a contribution due to surface tension:

$$\mathcal{G} = G_v(T,e)M_v + G_l(T,e)M_l + \gamma A, \tag{2.55}$$

where γ is the surface tension (or the surface energy per unit area). By conservation of mass $M_v = M_0 - M_l$, so, using equations (2.54), and (2.55) we get

$$\mathcal{G} - \mathcal{G}_0 = (G_l - G_v)M_l + \gamma A$$

and, using equation (2.53),

$$\mathcal{G} - \mathcal{G}_0 = -\frac{4}{3}\pi a^3 \rho_l R_v T \ln\left(\frac{e}{e_s(T)}\right) + 4\pi a^2 \gamma. \tag{2.56}$$

The variation with radius a of the total Gibbs free energy of the system therefore takes the form

$$\mathcal{G}(a) = \mathcal{G}_0 - \beta a^3 + \alpha a^2.$$

This is plotted in Figure 2.12, for two values of the relative humidity

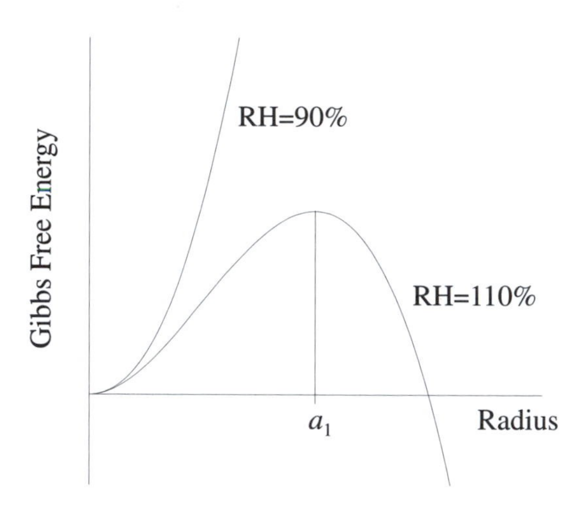

Figure 2.12 A schematic plot of the Gibbs free energy $\mathcal{G}$ as a function of the droplet radius a, for RH $= e/e_s =$ 90 and 110%.

e/e_s. Several useful facts can be learned from this figure. Note first that, if $e \le e_s(T)$ (indicating subsaturated or exactly saturated conditions), then the logarithm in equation (2.56) is negative or zero, so that $\beta \le 0$ and the curve of $\mathcal{G}(r)$ has no turning point other than $a = 0$. However, if $e > e_s(T)$ (indicating supersaturated conditions), then the logarithm is positive, $\beta > 0$, and there is a maximum of $\mathcal{G}$ at

$$a = a_1 = \frac{2\alpha}{3\beta} = \frac{2\gamma}{\rho_l R_v T \ln(e/e_s)}.$$

This is known as *Kelvin's formula*; it may also be written in the form

$$e = e_s(T) \exp\left(\frac{2\gamma}{\rho_l R_v T a_1}\right). \tag{2.57}$$

A well-known thermodynamic result is that a system at constant temperature and pressure tends to evolve in such a way that its Gibbs free energy decreases†; stable equilibrium is attained when the Gibbs free energy is a minimum. On the other hand, the point at which $\mathcal{G}$ is a maximum corresponds to an unstable equilibrium. Equation (2.57) shows how the partial pressure e of the vapour over a spherical droplet of radius a_1 in this equilibrium state differs from the SVP $e_s(T)$, in the presence of surface tension; see Figure 2.13. Note that a radius $a_1 = 0.01\,\mu\text{m}$ corresponds to a relative humidity RH $\simeq$ 112%, whereas $a_1 = 0.1\,\mu\text{m}$ corresponds to RH $\simeq$ 101%. As $a_1 \to \infty$ we recover the 'plane surface' result RH $\to$ 100%, i.e., $e \to e_s$.

If the droplet is in equilibrium at the radius a_1, condensation of a small quantity of vapour increases a slightly and the droplet then

† See also Section 6.1.

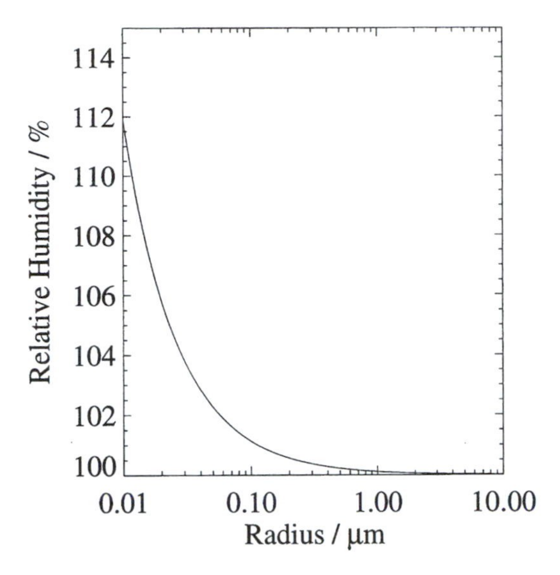

Figure 2.13 Plot of relative humidity RH $= e/e_s$ as a function of (unstable) equilibrium droplet radius a_1, at 5 °C.

continues to grow by further condensation. On the other hand, if a small amount of evaporation occurs at the equilibrium radius, the droplet continues to shrink by evaporation, eventually disappearing altogether.

We can therefore see that, if a cloud droplet is to survive, it must somehow attain a radius greater than a_1. Given that the relative humidity in clouds is seldom greater than about 101%, Figure 2.13 would require $a_1 \gtrsim 0.1\ \mu\text{m}$. A droplet of this size is unlikely to form by random collisions of smaller droplets, so a more viable process is for the droplet to condense on a small pre-existing solid or liquid particle, known as a *cloud-condensation nucleus.*

The analysis given above considers only the water vapour in the immediate neighbourhood of the droplet. However, if the droplet is to grow, there must be a continual supply of water vapour to its surface. This can happen by *diffusion* if there is a vapour density gradient in the region surrounding the droplet, with the vapour density increasing with distance. A simple representation of this diffusion is in terms of Fick's Law

$$\boldsymbol{f} = -D\nabla\rho_v,$$

where ρ_v is the vapour density, $\boldsymbol{f}$ is the vapour-flux vector and D is a diffusion coefficient, assumed constant. Assuming that at some instant the radius of the droplet is a and that the distribution of the vapour density is spherically symmetric, $\rho_v = \rho_v(r)$, the inward flux of mass of vapour through a sphere S_r of radius $r > a$ is

$$-\int_{S_r} \boldsymbol{f} \cdot \boldsymbol{n}\, dS = 4\pi r^2 \frac{d\rho_v}{dr} D,$$

where $\boldsymbol{n}$ is the outward normal to S_r; see Figure 2.14.

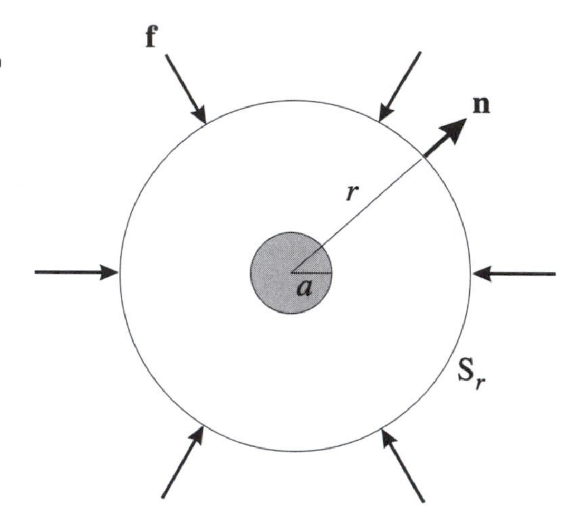

Figure 2.14 A sketch indicating inward diffusion of water vapour, through a sphere S_r, onto a droplet of radius a.

However, water vapour is lost only by condensation at $r = a$, so for $r > a$ this flux must be independent of r and equal to the rate of increase of mass of the droplet, dM_l/dt. Hence

$$\frac{d\rho_v}{dr} = \frac{dM_l/dt}{4\pi D}\frac{1}{r^2},$$

which can be integrated from $r = a$ to $r = \infty$ to give

$$\rho_v(a) = \rho_v(\infty) - \frac{1}{4\pi Da}\frac{dM_l}{dt}.$$

Therefore, using the ideal gas law $\rho_v = e/(R_v T)$ for the vapour,

$$\frac{dM_l}{dt} = 4\pi Da\,[\rho_v(\infty) - \rho_v(a)] = \frac{4\pi Da}{R_v T}\,[e(\infty) - e(a)]\,,$$

where $e(\infty)$ is the vapour pressure far from the droplet and $e(a)$ is the vapour pressure at its surface.

References

There are many basic textbooks covering the thermodynamics required in this chapter: examples include Adkins (1983), Finn (1993) and Mandl (1988). Good treatments of atmospheric thermodynamics, giving more detail than is presented here, are to be found in Wallace and Hobbs (1977), Salby (1996) and Bohren and Albrecht (1998). The original work on the concept of available potential energy was carried out by Lorenz (1955). More accurate empirical formulae for $e_s(T)$ than equation (2.38) are available: see, e.g., Bolton (1980). A comprehensive treatment of the tephigram is given by McIntosh and Thom (1983); see also McIlveen (1991). A good text on cloud physics is Rogers and Yau (1989).

Problems

2.1. Use the hydrostatic equation to show that the mass of a vertical column of air of unit cross-section, extending from the ground to a great height, is p_0/g, where p_0 is the surface pressure. Hence estimate the total mass of the atmosphere. Estimate also the total enthalpy of the atmosphere: this may be taken as a measure of the total heat content of the atmosphere.

The total mass of the oceans is 1.35×10^{21} kg. Estimate the total enthalpy of the ocean and calculate the depth of ocean that has the same enthalpy per unit horizontal area as the whole depth of the atmosphere. Comment on your result.

2.2. A balloon is required to carry an instrument payload of 100 kg to an altitude where the pressure is 12 hPa and the temperature is 230 K. Balloons are available with envelopes of non-stretch polythene with a thickness of 25 μm and density $10^3\,\mathrm{kg\,m^{-3}}$. What approximate radius of balloon is needed, assuming that after being charged with a measured 'bubble' of helium at launch it just becomes spherical at its floating altitude?

What are the relative merits and disadvantages of balloons, satellites and aircraft as instrument platforms?

2.3. Show that, if there is a uniform lapse rate $\Gamma = -dT/dz$, the pressure in the atmosphere is given by

$$p(z) = p_0 \left(1 - \frac{\Gamma z}{T_0}\right)^{g/(\Gamma R)}$$

where R is the gas constant per unit mass of air. Calculate the height at which the pressure is 0.1 of its surface value (p_0) assuming a surface temperature (T_0) of 290 K and (i) a uniform lapse rate of $10\,\mathrm{K\,km^{-1}}$ and (ii) a uniform temperature of 290 K.

2.4. Define the potential temperature θ. What is the temperature as a function of pressure in an atmosphere for which the lapse rate equals the DALR?

Show that an atmosphere of uniform potential temperature must have a finite depth. Calculate this depth for an Earth-like atmosphere with $\theta = 300$ K.

2.5. Derive an expression for the specific heat capacity c_p for air with a water vapour mass mixing ratio of μ. (Use the fact that $c_p = (\partial H/\partial T)_p$, where H is the enthalpy per unit mass; cf. equation (2.21).) Compare the values of the DALR for completely dry air and when $\mu = 0.02$. (The value of c_p for water vapour is given in Appendix A.)

2.6. Calculate the period of oscillation of an air parcel given that $dT/dz = -6.5\,\mathrm{K\,km^{-1}}$ and $T = 270$ K.

2.7. Show that the 'density lapse rate' satisfies

$$-\frac{g}{\rho}\frac{d\rho}{dz} = N^2 + \frac{g^2}{c_s^2},$$

where $c_s^2 = pc_p/(\rho c_v)$. ($c_s \approx 330\,\mathrm{m\,s^{-1}}$ is the speed of sound in air.) Comment on the implications for static stability if the density decreases very slowly with height.

2.8. Estimate the total potential energy per unit area for the Earth's atmosphere, assuming that it is at a uniform potential temperature of 300 K. An atmosphere similar to that of the Earth is initially all at a uniform potential temperature θ. The air in one hemisphere is heated is such a way that its potential temperature is raised uniformly by an amount $\Delta\theta$. Find an expression for the amount of *available* potential energy per unit area that is generated.

2.9. Estimate the terms $L\mu_s/(RT)$ and $L^2\mu_s/(c_pR_vT^2)$ in equation (2.48) for the SALR, given a reasonable estimate of T and p, using Figure 2.7 to estimate μ_s. Hence derive an estimate for the SALR.

2.10. Plot the measurements in Table 2.2 on a tephigram chart, using straight-line segments between data points, and answer the following questions.

(*a*) What is the pressure at the tropopause?

(*b*) What parts of the ascent are stable for dry air and for saturated air?

(*c*) What is the mass mixing ratio of water vapour at 1000 hPa and at 500 hPa?

(*d*) The night air is clear and the surface cools radiatively, resulting in a temperature inversion near the ground. How many degrees of cooling are required near the ground for fog to begin to form?

(*e*) The rising Sun disperses the fog and heats the ground, destroying the inversion. After further heating a parcel of air rises adiabatically. At what (pressure) level will condensation occur?

(*f*) What will be the level of the top of the convective clouds which develop?
(From Houghton (1986).)

2.11. Consider the layer of air initially between 780 and 700 hPa in the Liverpool midnight ascent. Suppose it is pushed up (adiabatically) by a wedge of colder air (a cold front) driving in underneath it so that the pressure at the bottom of the layer is reduced to 680 hPa. What is the new pressure at the top of the layer? Use the tephigram chart to find the new temperatures at the bottom and top of the layer. Has the

Table 2.2. *Radiosonde measurements made at midnight in June during an ascent from Liverpool.*

Pressure/hPa	Temperature/°C	Dew (or frost) point/°C
1000	13	11
940	9.5	8
900	7	5
780	0	−3
700	−5	−11
600	−11	−17
500	−20	−28
400	−32	−42
300	−47	
200	−49	

stability of parcels within the layer with respect to wet ascent been changed by the layer being lifted?

Such a situation is called one of *potential instability* or *convective instability*; the amount of such instability will determine how much convective development will occur.

Carry out the same exercise for the layer between 900 and 780 hPa.

(From Houghton (1986).)

2.12. Show that no energy is needed or released if a moist parcel rises adiabatically and quasi-statically in an atmosphere which matches the dry and then the saturated adiabatics of the parcel.

Draw the dry adiabatic, followed by the wet adiabatic, for a parcel originating at the surface of the Liverpool ascent and note the two points of intersection with the ascent. Show that energy is needed to raise the parcel adiabatically in the atmosphere to the first point of intersection and that energy is released if the parcel rises from the first to the second point. Calculate these energies for 1 kg of air and show that the net energy released is positive.

Such a condition is known as *latent instability*. Once the surface has been heated enough to initiate vertical motion, energy can be released and intense convective activity (and perhaps thunderstorms) may develop.

(From Houghton (1986).)

2.13. Calculate the times taken for water drops of radii 1, 10 and 100 μm to fall a distance of 1 km in air at the terminal velocity (take the dynamic viscosity η for air to be $1.7 \times 10^{-5}\,\mathrm{kg\,m^{-1}s^{-1}}$). Use Stokes' Law, which states that the viscous force on a spherical drop of radius r and speed v is $6\pi\eta r v$.

The condition for Stokes' Law to hold is that the dimensionless *Reynolds Number*, $\mathrm{Re} \equiv \rho v r/\eta$, should be small, where ρ is the air density (see, e.g., Acheson (1990)). Check the validity of this condition for each drop radius and comment on your result.

2.14. Show that the temperature T_i at the surface of a spherical ice crystal of mass M_i growing in a cloud of water droplets at temperature T is given by

$$T_i - T = \frac{L_s}{4\pi\lambda}\frac{\dot{M}_i}{r},$$

where L_s is the specific latent heat of sublimation of ice, $\dot{M}_i$ is the rate of the increase in mass, r is the radius and λ is the thermal conductivity of the air.

Show also that

$$\frac{\dot{M}_i}{r} = \frac{4\pi D}{R_v}\left(\frac{e}{T} - \frac{e_i}{T_i}\right),$$

where D is the diffusion coefficient of water molecules in air, R_v is the gas constant per unit mass of water vapour and e and e_i are the saturation vapour pressures over water and ice respectively.

Given that $(e/T) - (e_i/T_i) = 6.7 \times 10^{-2}\,\mathrm{Pa\,K^{-1}}$ at the relevant temperature ($\approx -10\,°\mathrm{C}$) calculate (i) the temperature difference between the surface of the ice crystal and the air some distance away and (ii) the time taken for the crystal to grow from $1\,\mu\mathrm{m}$ to $100\,\mu\mathrm{m}$ radius.
(Take $\lambda = 2.4 \times 10^{-2}\,\mathrm{W\,m^{-1}\,K^{-1}}$ and $D = 0.23 \times 10^{-4}\,\mathrm{m^2\,s^{-1}}$.)

CHAPTER THREE

Atmospheric radiation

This chapter describes some aspects of energy transfer by electromagnetic radiation in the atmosphere. In Section 3.1 we introduce the Planck function, the solar spectrum and the concept of local thermodynamic equilibrium. In Section 3.2 we list some formal definitions of radiometric quantities and then derive and solve the *radiative-transfer equation*, which describes the way in which radiative power is affected by extinction and emission of radiation. In Section 3.3 we present some basic aspects of molecular spectrosopy and give some of the properties of spectral line shapes. In Section 3.4 we introduce the concept of *transmittance*, the fraction of radiative power that survives propagation from one point to another. In Section 3.5 we consider the absorption and emission of infra-red radiation and the absorption of ultra-violet radiation by gases in the atmosphere. This absorption and emission lead to heating and cooling; the principles of the calculation of heating rates are outlined in Section 3.6. In Section 3.7, we revisit the greenhouse effect, investigating a more realistic model than that described in Section 1.3.2. Finally, in Section 3.8, we discuss a simple model of atmospheric scattering.

The solution of the radiative transfer equation also plays an important role in certain aspects of atmospheric remote sounding. This will be covered in Chapter 7.

It is an unfortunate fact that quantitative calculations of radiative heating rates, for example, involve considerable geometric and algebraic detail, which tend to distract attention from the basic physics of the processes. This chapter emphasises the underlying physical concepts as much as possible and keeps the formal mathematics to a minimum; however, the mathematics cannot be avoided entirely.

3.1 Basic physical concepts

The subject of atmospheric radiation is concerned with the transfer of energy within the atmosphere by photons, or equivalently by electromagnetic waves. The relevant photons fall into two classes.

- Solar (or short-wave) photons, emitted by the Sun; these correspond to ultra-violet, visible and infra-red[†] wavelengths between about 0.1 and 4 μm.
- Thermal (or long-wave) photons, emitted by the atmosphere or the Earth's surface; these correspond mainly to infra-red wavelengths, between about 4 and 100 μm.

These two wavelength ranges represent spectral regions of significant black-body emission at temperatures of about 6000 K (a temperature representative of the solar photosphere) and 288 K (the Earth's mean surface temperature), respectively; see Section 3.1.1.

To study the effects of atmospheric radiation, it is necessary to investigate the interaction between photons and atmospheric gases. One way in which solar photons may be lost is by interaction with molecules at certain discrete frequencies, each frequency ν corresponding to an orbital transition of an electron to a higher energy level according to the formula $\Delta E = h\nu$, where ΔE is the difference in energy levels and h is Planck's constant. (The corresponding wavelength λ is given by $\lambda = c/\nu = hc/\Delta E$, where c is the speed of light.) However, the resulting excited state has a limited lifetime and the excitation energy may be lost again in one of two ways.

(a) The electron falls back to the ground state, re-emitting a photon of the same energy and frequency as the original photon, but in a random direction. This process is called *radiative decay*.

(b) At sufficiently high pressures, molecular collisions are likely to occur before re-emission takes place, leading to transfer of the excitation energy ΔE to other forms of energy: the photon is then said to have been *absorbed*. If kinetic energy is produced in the process, this will quickly be shared between molecules by collisional interactions and (since thermal energy is the macroscopic expression of molecular kinetic energy) local *heating* of the atmosphere takes place. This transfer of photon energy to heat is called *thermalization* or *quenching*.

Radiative decay, defined in (a), is an example of the *scattering* of a photon of a given, discrete, frequency by an atmospheric

[†] Recall that the wavelength of visible light lies between about 0.4 μm = 400 nm (violet) and about 0.7 μm = 700 nm (red), with ultra-violet shorter than 400 nm and infra-red longer than 0.7 μm.

molecule. More important for atmospheric physics, however, are the *continuum* processes of the scattering of photons, over broad ranges of frequencies, by atmospheric molecules and by solid or liquid particles in suspension in the atmosphere. (A suspension of this kind is called an *aerosol.*) In the case of scattering by molecules, whose dimensions are much less than the wavelength of the solar radiation, we have *Rayleigh scattering* (see Section 7.3.2 for more details). In the case of scattering by aerosol particles such as dust and smog, whose dimensions are comparable to the wavelength of the solar radiation, we have the more complex *Mie scattering*. (For scattering by particles such as cloud droplets or raindrops, which are much larger than the wavelength of the solar radiation, geometric optics applies; this describes optical phenomena such as rainbows and haloes.)

The term *extinction* is used to denote loss of energy from an incoming photon. This can occur either by absorption or by scattering.

Absorption of solar photons may also cause *photo-dissociation*, i.e., the breakdown of the molecules, leading to photochemical reactions, and *photo-ionisation*, in which outer electrons are stripped from atoms. These interactions occur over a continuous range of frequencies, provided that the energy of the incoming photon is large enough.

Thermal photons may be absorbed and scattered in a similar manner to solar photons. They may also be *emitted*, by the inverse process to absorption, with energy being drawn from molecular kinetic energy, thus leading to a local cooling of the atmosphere. However, at the infra-red frequencies involved here, the relevant ΔE corresponds to a difference between the energies of pairs of vibrationally or rotationally excited states of the emitting molecule, rather than the energy of an electronic transition; see Section 3.3.

In this chapter we shall use the frequency ν and wavelength λ, as appropriate, to describe radiative properties. A related quantity is the wavenumber $\tilde{\nu} = \nu/c = 1/\lambda$; this is commonly used in spectroscopy and is often measured in units of cm^{-1}. It should not be confused with the quantity $k = 2\pi/\lambda$, also called the wavenumber, that is used in the description of wave motion (e.g., Section 5.4).

3.1.1 The Planck function

An *isothermal cavity* is defined as a cavity whose walls are maintained at a uniform temperature. Under *thermodynamic equilibrium* conditions, the radiation within such a cavity is in equilibrium with the cavity walls and it can be shown that the spectral energy density (the energy per unit volume per unit frequency interval) depends only on frequency and temperature; the radiation is also isotropic.

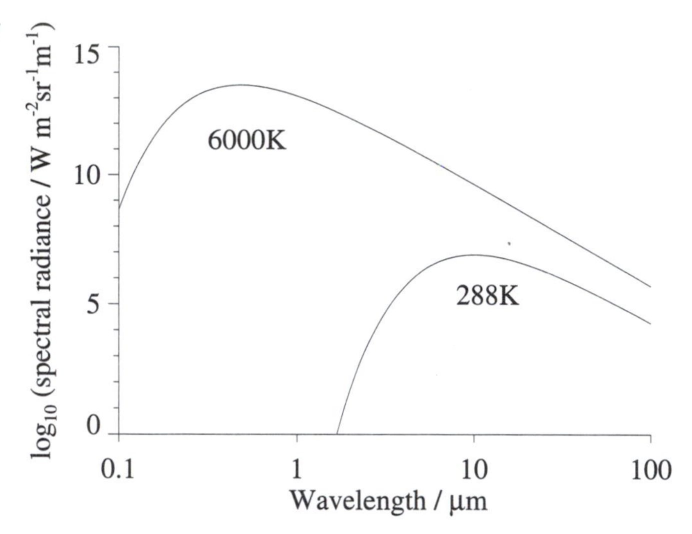

Figure 3.1 Logarithm of the black-body spectral radiance $B_\lambda(T)$, plotted against the logarithm of wavelength λ, for T = 6000 K, a typical temperature of the solar photosphere, and 288 K, the Earth's mean surface temperature.

If a small hole is cut in the cavity then the emitted radiation will have the same form as the radiation within the cavity. Radiation of this kind is called *black-body radiation.*

Planck's Law states that the spectral energy density of black-body radiation at temperature T is given by

$$u_\nu(T) = \frac{8\pi h\nu^3}{c^3\{\exp[h\nu/(kT)] - 1\}},$$

where k is Boltzmann's constant. Since the photons carrying this energy are moving isotropically, the energy density associated with the group of photons moving within a small solid angle $\Delta\Omega$ steradians is $u_\nu\,\Delta\Omega/(4\pi)$. Consideration of the energy flow per unit time, per unit area, transferred at speed c by this group of photons then shows that the power per unit area, per unit solid angle, per unit frequency interval (the spectral radiance; see Section 3.2.1) for black-body radiation at temperature T is

$$B_\nu(T) = \frac{2h\nu^3}{c^2\{\exp[h\nu/(kT)] - 1\}}; \tag{3.1}$$

this is called the *Planck function.*

The black-body spectral radiance can also be written in terms of the power per unit area, per unit solid angle, per unit *wavelength* interval,

$$B_\lambda(T) = \frac{2hc^2}{\lambda^5\{\exp[hc/(\lambda kT)] - 1\}}; \tag{3.2}$$

see Problem 3.1. Figure 3.1 shows B_λ for temperatures of 6000 and 288 K. Since large ranges of wavelengths and spectral radiances are under consideration, it is convenient to use a log–log plot here.

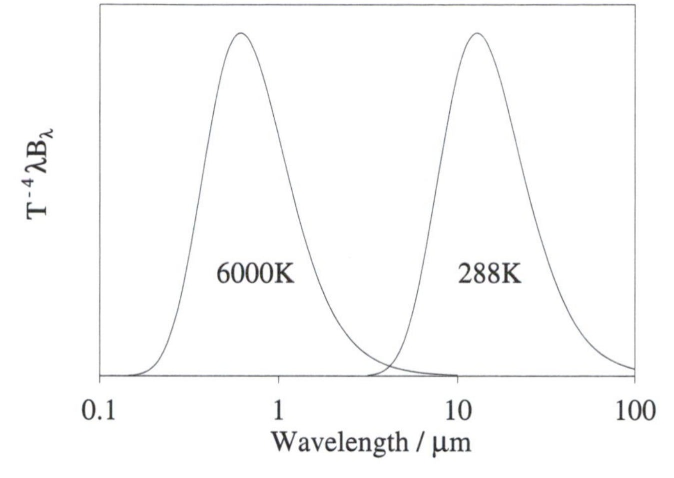

Figure 3.2 The black-body spectral radiance $B_\lambda(T)$, multiplied by $T^{-4}\lambda$, plotted against the logarithm of wavelength λ, for $T =$ 6000 and 288 K.

Note that the curve for 6000 K lies above that for 288 K for all wavelengths (but see also Problem 3.3).

If B_λ is integrated over all wavelengths, we obtain the *black-body radiance*

$$\int_0^\infty B_\lambda(T)\,d\lambda = \frac{\sigma}{\pi}T^4, \tag{3.3}$$

where σ is the Stefan–Boltzmann constant. In terms of an integral over $\ln(\lambda)$, this gives

$$T^{-4}\int_{-\infty}^{\infty} \lambda B_\lambda(T)\,d(\ln\lambda) = \frac{\sigma}{\pi}.$$

This suggests plotting $T^{-4}\lambda B_\lambda$ against $\ln\lambda$: the area under the resulting curve is then independent of T. Curves of this kind are shown in Figure 3.2. Note that, with this normalisation, there is little overlap between the black-body spectral radiances at 6000 and 288 K.

A *black body* is defined as a body that completely absorbs all radiation falling on it. It can be shown that the radiation emitted by a black body is black-body radiation, as defined above. The concept of a black body is an idealisation: a real body will emit less radiation than this. The *spectral emittance* ϵ_ν of a body is the ratio of the spectral radiance from that body to the spectral radiance from a black body; therefore $\epsilon_\nu \leq 1$. It follows that a black body emits the maximum possible amount of energy in each frequency interval, at a given temperature. We can also define the *spectral absorptance* α_ν as the fraction of energy per unit frequency interval falling on a body that is absorbed. *Kirchhoff's Law* states that $\epsilon_\nu = \alpha_\nu$; i.e., at a given temperature and frequency the spectral emittance of a body equals its spectral absorptance.

3.1.2 Local thermodynamic equilibrium

The standard derivation of the Planck function (3.1) applies to an isothermal cavity containing radiation, but not containing matter. On the other hand, statistical mechanics shows that the energy levels of a material system (for example a gas) in equilibrium at temperature T will be populated according to the *Boltzmann distribution*, when radiation is neglected. That is, the numbers n_1 and n_2 of molecules in states of energy E_1 and E_2 and with statistical weights (or degeneracies) g_1 and g_2, respectively, are in the ratio given by the Boltzmann distribution

$$\frac{n_1}{n_2} = \frac{g_1}{g_2} e^{-(E_1-E_2)/(kT)}. \tag{3.4}$$

In the case of a gas, this equilibrium ratio is maintained by collisions between the gas molecules[†].

When matter and radiation are both contained in an isothermal cavity and the interaction between the matter and radiation is sufficiently *weak*, then in thermodynamic equilibrium the radiation will continue to satisfy Planck's Law and the matter will continue to satisfy the Boltzmann distribution. The interaction between the matter and the radiation is essential to bring about thermodynamic equilibrium, but it must not be so strong as to lead to significant departures of the radiation from Planck's Law or of the matter from the Boltzmann distribution[‡]. The interaction will be sufficiently weak, for a given pair of energy levels, if the mean time between collisions for a given molecule, τ_c say (which is inversely proportional to the pressure p), is much shorter than the lifetime for radiative decay, τ_d say, for the given levels. This state of thermodynamic equilibrium will therefore hold if the pressure is large enough, for a given transition between energy states.

The atmosphere does not have a uniform temperature, so we cannot regard it as being in strict thermodynamic equilibrium. However, at high enough pressures, molecular collisions are sufficiently rapid for Boltzmann's equation (3.4) to hold for each small portion of the atmosphere, given the local value of T. Such a portion of the atmosphere is said to be in *local thermodynamic equilibrium* (LTE) with respect to the given energy states. We can then, for example, use the Planck function (3.1) – which is derived under thermodynamic equilibrium conditions – to represent the spectral radiance.

[†] For an ideal gas, these collisions must be infrequent; that is, the interaction between the molecules must be weak. However, if equilibrium is to be maintained, some such interaction is essential. The necessity of a weak interaction between different 'aspects' of a system for maintenance of equilibrium is a familiar one in statistical mechanics; see Mandl (1988), page 51.

[‡] For an advanced treatment, see Landau and Lifshitz (1980), Chapters IV and V.

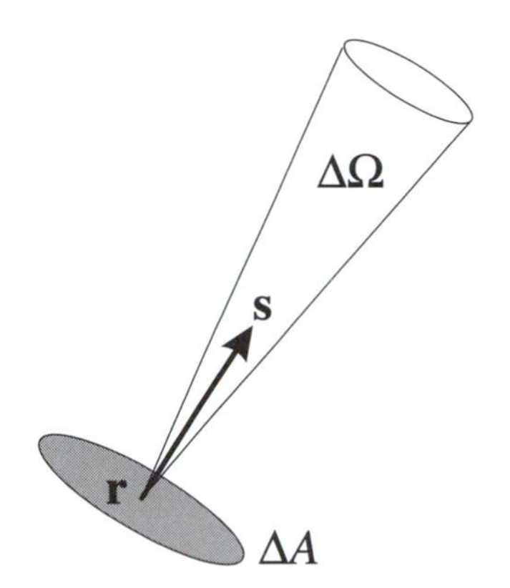

Figure 3.3 Illustrating the definition of spectral radiance; see the text. Although a cone is shown only emerging from the point $\boldsymbol{r}$, we must also consider parallel cones emerging from the other points in the area ΔA.

It can be shown that LTE applies to translational modes (those associated with molecular kinetic energy and macroscopic thermal energy) below about 500 km altitude. For the vibrational and rotational modes involved in the absorption and emission from most radiatively active gases, LTE holds for pressures greater than about 0.1 hPa, corresponding roughly to altitudes below about 60 km. Methods for calculating the spectral radiance when LTE does not hold are complex and will not be discussed in this book.

3.2 The radiative-transfer equation

3.2.1 Radiometric quantities

Several different, but related, quantities are used in the description and measurement of radiation. The most important are as follows.

- The *spectral radiance* (or monochromatic radiance) $L_\nu(\boldsymbol{r}, \boldsymbol{s})$ is the power per unit area, per unit solid angle, per unit frequency interval in the neighbourhood of the frequency ν, at a point $\boldsymbol{r}$, in the direction of the unit vector $\boldsymbol{s}$. It is measured in W m^{-2} steradian^{-1} Hz^{-1}. The spectral radiance can be visualised in terms of the photons emerging from a small area ΔA with unit normal $\boldsymbol{s}$, centred at a point $\boldsymbol{r}$: see Figure 3.3. Consider those photons whose momentum vectors lie within a cone of small solid angle $\Delta\Omega$ centred on the direction $\boldsymbol{s}$ and whose frequencies lie between ν and $\nu + \Delta\nu$. Then $L_\nu \, \Delta A \, \Delta\Omega \, \Delta\nu$ is the energy transferred by these photons, per unit time, from 'below ' the area ΔA to 'above'. (Here 'below' means in the direction $-\boldsymbol{s}$ and 'above' means in the direction $\boldsymbol{s}$.)

  We have already encountered a special case of the spectral radiance, for isotropic black-body radiation in an isothermal cavity (Section 3.1.1), when $L_\nu = B_\nu(T)$, the Planck function;

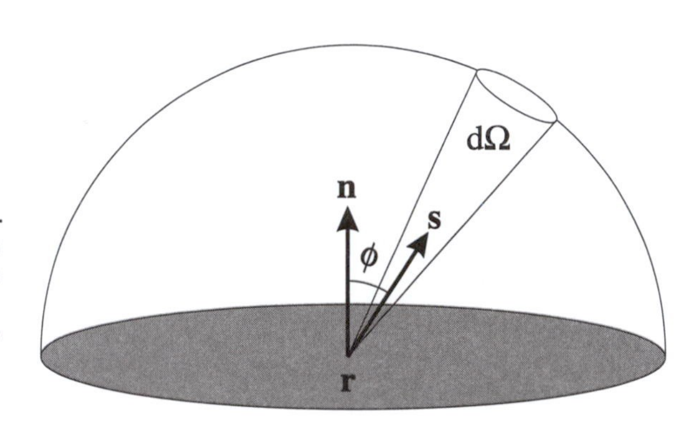

Figure 3.4 Illustrating the calculation of the spectral irradiance by integrating the spectral radiance over the hemisphere above the shaded horizontal surface. The scalar product $\boldsymbol{n} \cdot \boldsymbol{s}$ in equation (3.5) arises from the projection of the unit area perpendicular to $\boldsymbol{s}$ in the direction $\boldsymbol{n}$ of the normal.

this depends only on the temperature of the cavity and is independent of position and direction.

- The *radiance* $L(\boldsymbol{r}, \boldsymbol{s})$ is the power per unit area, per unit solid angle at a point $\boldsymbol{r}$ in the direction of the unit vector $\boldsymbol{s}$; in other words it is the integral of L_ν over frequency:

$$L(\boldsymbol{r}, \boldsymbol{s}) = \int_0^\infty L_\nu(\boldsymbol{r}, \boldsymbol{s})\, d\nu.$$

Its units are $\mathrm{W\,m^{-2}\,steradian^{-1}}$.

- The *spectral irradiance* (or monochromatic irradiance) $F_\nu(\boldsymbol{r}, \boldsymbol{n})$ is the power per unit area, per unit frequency interval in the neighbourhood of the frequency ν, at a point $\boldsymbol{r}$ through a surface of normal $\boldsymbol{n}$; its units are $\mathrm{W\,m^{-2}\,Hz^{-1}}$. It is obtained from the spectral radiance by integration over a hemisphere on one side of the surface:

$$F_\nu(\boldsymbol{r}, \boldsymbol{n}) = \int_{2\pi} L_\nu(\boldsymbol{r}, \boldsymbol{s}) \boldsymbol{n} \cdot \boldsymbol{s}\, d\Omega(\boldsymbol{s}), \tag{3.5}$$

where $d\Omega(\boldsymbol{s})$ is the element of solid angle in the direction $\boldsymbol{s}$; see Figure 3.4. We therefore integrate over *all* photons in the frequency interval that emerge into the region above the surface†. As with the Planck function, the spectral radiance and spectral irradiance can alternatively be expressed per unit *wavelength* interval.

- The *irradiance* (or flux density) $F(\boldsymbol{r}, \boldsymbol{n})$ is the power per unit area at a point $\boldsymbol{r}$ through a surface of normal $\boldsymbol{n}$, i.e., the integral of F_ν over frequency, and also the integral of the

† The angular integration is similar to that used in the kinetic theory of gases; indeed many of the geometric concepts that apply to moving molecules in kinetic theory apply equally to photons, with the simplification that the photons all have the same speed c, rather than a distribution of speeds.

radiance L over a hemisphere:

$$F(\boldsymbol{r},\boldsymbol{n}) = \int_0^\infty F_\nu(\boldsymbol{r},\boldsymbol{n})\,d\nu = \int_{2\pi} L(\boldsymbol{r},\boldsymbol{s})\boldsymbol{n}\cdot\boldsymbol{s}\,d\Omega(\boldsymbol{s}).$$

Its units are $\mathrm{W\,m^{-2}}$.

It must be borne in mind that the irradiance has a specific direction associated with it; for example, if the surface in question (assumed for the present argument to be an imaginary, rather than a material, surface) is horizontal and the normal $\boldsymbol{n}$ points upwards, then the irradiance under consideration (denoted by $F^\uparrow$) is associated with upward-moving photons. Conversely, the irradiance $F^\downarrow = F(\boldsymbol{r},-\boldsymbol{n})$ is associated with downward-moving photons. If we require the *net* upward power per unit area, F_z say, then we must take the difference:

$$F_z = F^\uparrow - F^\downarrow.$$

In terms of electromagnetic quantities, this net upward irradiance equals the vertical component of the Poynting vector.

There is a simple relationship between the radiance and irradiance from an isothermal plane surface, at temperature T, that emits black-body radiation. Since the black-body radiation is isotropic, $L_\nu = B_\nu(T)$ is independent of $\boldsymbol{s}$ and $\boldsymbol{r}$, so the hemispheric integral is straightforward. Equation (3.5) for the spectral irradiance becomes

$$F_\nu(\boldsymbol{r},\boldsymbol{n}) = \int_{2\pi} L_\nu\boldsymbol{n}\cdot\boldsymbol{s}\,d\Omega(\boldsymbol{s}) = 2\pi B_\nu \int_0^{\pi/2} \cos\phi\sin\phi\,d\phi = \pi B_\nu(T), \tag{3.6}$$

where ϕ is the angle between $\boldsymbol{s}$ and the normal $\boldsymbol{n}$ (see Figure 3.4) so that $d\Omega = 2\pi\sin\phi\,d\phi$, since there is axisymmetry around the normal. Integrating over all ν we obtain the Stefan–Boltzmann law for the irradiance

$$F(\boldsymbol{r},\boldsymbol{n}) = \pi\int_0^\infty B_\nu(T)\,d\nu = \sigma T^4, \tag{3.7}$$

using the frequency-integral analogue of equation (3.3).

3.2.2 Extinction and emission

Consider a beam of radiation of unit cross-sectional area, moving in a small range of solid angles $\Delta\Omega$ about the direction $\boldsymbol{s}$; see Figure 3.5. If the photons experience absorption or scattering in a small distance ds along the beam, due to the presence of a radiatively active gas†, then the spectral radiance L_ν will be reduced.

† We refer to gases here, but the same considerations also apply to a gas containing a suspension of solid particles or liquid droplets.

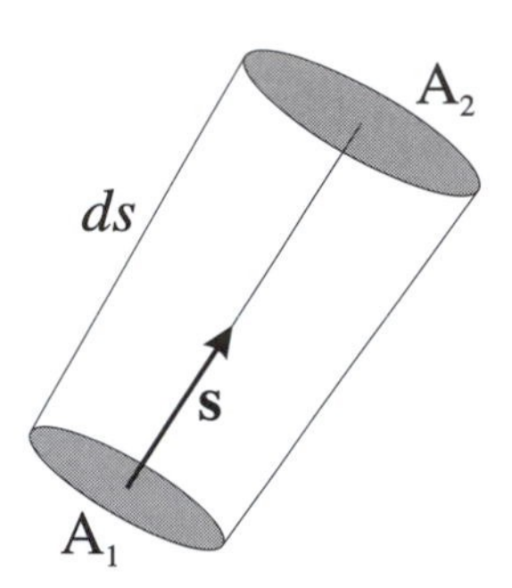

Figure 3.5 A beam, or 'pencil', of radiation travelling a distance ds from surface A_1 of unit area to a surface A_2. (Scattering of photons out of or into the beam is not indicated.) The area of A_2 is slightly greater than unity, owing to the divergence of photons into a solid angle $\Delta\Omega$ from each point of A_1; see Figure 3.3. However, to leading order the volume of the pencil is still given by (unit area) $\times\, ds$.

The physics of the process is complex; however, it may be summed up by *Lambert's Law*, which states that the fractional decrease of the spectral radiance is proportional to the mass of absorbing or scattering material encountered by the beam in a distance ds. Since the beam has unit cross-sectional area, this mass is $\rho_a\, ds$, where ρ_a is the density of the radiatively active gas, so

$$dL_\nu = -k_\nu(s)\rho_a(s)L_\nu(s)\, ds. \tag{3.8}$$

(The dependence of L_ν on the direction vector $\boldsymbol{s}$ is omitted here for clarity.) The quantity k_ν is called the *extinction coefficient*; it is the sum of an absorption coefficient a_ν and a scattering coefficient s_ν, defined in an obvious manner in terms of the contributions to dL_ν from absorption and scattering, respectively:

$$k_\nu = a_\nu + s_\nu. \tag{3.9}$$

The extinction coefficient k_ν generally depends on temperature and pressure, and can be regarded as calculable from detailed quantum mechanics or as an empirical quantity, to be derived from measurements. (Note that ρ_a is *not* generally the same as the total gas density ρ.)

If the gas is also emitting photons of frequency ν, an extra term must be added to the right-hand side of equation (3.8), to represent the additional power per unit area introduced into the beam. This term will also be proportional to the mass $\rho_a\, ds$, so it is convenient to write it as $k_\nu \rho_a J_\nu\, ds$, where $J_\nu(s)$ is called the *source function.* Including both extinction and emission we therefore obtain the *radiative-transfer equation* (also called *Schwarzschild's equation*)

$$\frac{dL_\nu}{ds} = -k_\nu \rho_a \big(L_\nu - J_\nu\big). \tag{3.10}$$

If k_ν, ρ_a and J_ν are given as functions of distance s, a formal solution of the radiative-transfer equation can be obtained as follows. First introduce the *optical path* χ_ν, defined by

$$\chi_\nu(s) = \int_{s_0}^{s} k_\nu(s')\rho_a(s')\, ds', \tag{3.11}$$

where s_0 is the start of the path; then equation (3.10) can be written as

$$\frac{dL_\nu}{d\chi_\nu} + L_\nu = J_\nu. \tag{3.12}$$

Using the integrating factor $\exp(\chi_\nu)$ we can integrate equation (3.12) to get

$$L_\nu e^{\chi_\nu} = \int J_\nu e^{\chi_\nu{}'} \, d\chi_\nu' + \text{constant}.$$

If the spectral radiance equals $L_{\nu 0}$ at the point s_0 then

$$L_\nu(s) = \int_0^{\chi_\nu} J_\nu(\chi') e^{-(\chi_\nu - \chi')} \, d\chi' + L_{\nu 0} e^{-\chi_\nu}. \tag{3.13}$$

Note that, in the absence of emission ($J_\nu = 0$), the spectral radiance falls exponentially†, decreasing by a factor of e over a distance corresponding to unit optical path. A region is said to be *optically thick* at a frequency ν if the total optical path χ_ν through the region is greater than 1 and *optically thin* if the total optical path is less than 1. A photon is likely to be absorbed or scattered within an optically thick region, but is likely to traverse an optically thin region without absorption or scattering.

As a simple example, suppose that the extinction coefficient k_ν and the density ρ_a of the radiatively active gas are both constant and take $s_0 = 0$. Then from equation (3.11) the optical path is proportional to the distance s, $\chi_\nu = k_\nu \rho_a s$, so

$$L_\nu(s) = k_\nu \rho_a \int_0^s J_\nu(s') e^{-k_\nu \rho_a (s - s')} \, ds' + L_{\nu 0} e^{-k_\nu \rho_a s}. \tag{3.14}$$

The radiance $L_\nu(s)$ reaching s thus has a simple interpretation, as follows. The second term on the right-hand side of equation (3.14) represents the radiance at the starting point $s = 0$, attenuated by an exponential factor due to extinction over the distance s, while the integral represents the sum of contributions emitted from elements ds' at different distances s' along the path, each attenuated by the factor $\exp[-k_\nu \rho_a (s - s')]$ due to extinction over the remaining distance $s - s'$. (See Figure 3.6.)

Under local thermodynamic equilibrium conditions (see Section 3.1.2), in the absence of scattering, the source function equals the black-body spectral radiance, or the Planck function, (3.1). This can be shown by using Kirchhoff's Law (see Section 3.1.1), which holds under LTE conditions, as follows. The radiance emitted from a mass $\rho_a \, ds$ of gas in the beam is $k_\nu \rho_a \, ds \, J_\nu$ and the radiance absorbed is $k_\nu \rho_a \, ds \, L_\nu$; cf. equation (3.10). Hence the spectral emittance, the ratio of the emitted radiance to the radiance emitted by a black body, is $\epsilon_\nu = k_\nu \rho_a \, ds \, J_\nu / B_\nu$. The spectral absorptance, the fraction of

† This exponential decay is known as Beer's Law.

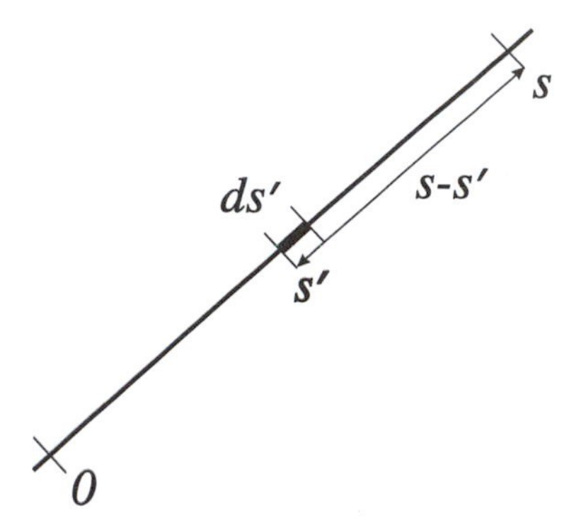

Figure 3.6 Illustrating the terms in equation (3.14).

incident radiance that is absorbed, is $\alpha_\nu = k_\nu \rho_a \, ds \, L_\nu / L_\nu = k_\nu \rho_a \, ds$, neglecting scattering. However, Kirchhoff's Law states that $\epsilon_\nu = \alpha_\nu$ and hence $J_\nu = B_\nu$.

3.2.3 The diffuse approximation

In radiative calculations we can often assume that the properties of the atmosphere and the radiation depend only on the vertical coordinate z†. This is the *plane-parallel atmosphere* assumption and we shall make it from now on. The net irradiance (taking account of photons moving in opposite directions) is then vertical and equal to $F_z(z)$ defined in Section 3.2.1. Calculation of the upward and downward irradiances involves the integration, over the solid angle, of radiances such as that in equation (3.13), with the direction vector $\boldsymbol{s}$ re-inserted. This integration, though conceptually straightforward, is quite complicated in practice. However, a simple and surprisingly accurate alternative is to use the *diffuse approximation*. This states that we can replace the radiative transfer equation (3.12), for calculating the spectral radiances L_ν along a set of slanting downward paths (with $J_\nu = B_\nu$, assuming LTE conditions), by the single equation

$$\frac{dF_\nu^\downarrow}{d\chi_\nu^*} + F_\nu^\downarrow = \pi B_\nu \qquad (3.15)$$

for the downward spectral irradiance $F_\nu^\downarrow$ along a *vertical* path, where

$$\chi_\nu^* \approx 1.66\chi_\nu. \qquad (3.16)$$

The quantity χ_ν is the optical path measured downwards from the top of the atmosphere and is called the *optical depth*; see equation (3.29). The quantity χ_ν^* is a scaled optical depth. The

† An important exception is when horizontal inhomogeneities due to clouds are present.

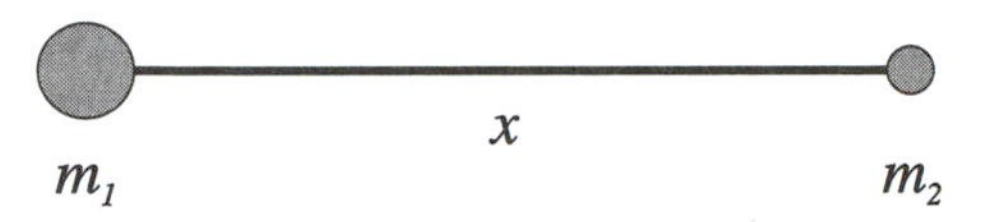

Figure 3.7 A simple representation of a diatomic molecule, composed of atoms of masses m_1 and m_2, separated by a distance x.

factor π on the right-hand side of equation (3.15) arises from the calculation of the black-body irradiance, as in equation (3.6). A similar equation, but with a change in sign of χ_ν^*, holds for the upward spectral irradiance; see equation (3.36a).

3.3 Basic spectroscopy of molecules

3.3.1 Vibrational and rotational states

It was mentioned in Section 3.1 that the absorption and emission of thermal photons depend on energy differences between vibrationally or rotationally excited states of atmospheric molecules. The detailed spectroscopy of atmospheric molecules is a highly complex subject, well beyond the scope of this book. However, the basic principles are comparatively simple and will be sketched out in this section. As an illustration, we consider the vibrational and rotational states of a diatomic molecule, composed of an atom of mass m_1 and an atom of mass m_2; see Figure 3.7.

The classical picture of the vibration of this molecule assumes that the atoms are bound by a light 'spring', of spring constant K, that resists deviations of the distance x between the atoms from its equilibrium value, x_0 say. It is a straightforward exercise in classical mechanics to show that the frequency of oscillation of the system is ν_0, given by

$$2\pi\nu_0 = \left(\frac{K}{m_r}\right)^{1/2}, \qquad \text{where} \qquad m_r = \frac{m_1 m_2}{m_1 + m_2}.$$

m_r is called the *reduced mass*. The quantum-mechanical theory of the harmonic oscillator requires that we insert the potential function

$$V(x) = \tfrac{1}{2}K(x - x_0)^2$$

for this system into Schrödinger's equation; there results an infinite set of energy levels, given by

$$E_v = h\nu_0\left(v + \tfrac{1}{2}\right), \qquad \text{where} \qquad v = 0, 1, 2, \ldots. \tag{3.17}$$

Here v is the *vibrational quantum number* and takes integer values. The levels are *non-degenerate*; that is, there is only one state corresponding to each energy value E_v. The quantum-mechanical *selection rule* for this system states that, in a transition from one state to another, associated with dipole radiation, v can only increase or decrease by unity ($\Delta v = \pm 1$), corresponding to an energy change $\Delta E = \pm h\nu_0$.

We now consider the rotation of the same diatomic molecule, ignoring vibrations so that $x = x_0$. Classically, the molecule has a moment of inertia $I = m_r x_0$ and can rotate with any angular momentum. The quantum-mechanical analysis of the system is similar to that for the angular structure of the hydrogen atom, for example. It is found that the only allowed values of the squared angular momentum are $\hbar^2 J(J+1)$ where J is an integer, the *rotational quantum number*, and $\hbar = h/(2\pi)$. The corresponding energy levels are given by

$$E_J = \frac{1}{2I} J(J+1)\hbar^2, \qquad \text{where} \qquad J = 0, 1, 2, \ldots ; \tag{3.18}$$

these levels are *degenerate*, with $2J+1$ states corresponding to each energy value E_J. The selection rule for dipole radiation in this case is $\Delta J = \pm 1$.

Equations (3.17) and (3.18) illustrate the fact that, when quantum mechanics is taken into account, only discrete (quantized) energy levels are allowed for vibrational and rotational states of the simple diatomic molecule. Furthermore, transitions between these quantized levels, associated with absorption or emission of photons, must lead to energy differences ΔE and hence photon frequencies $\nu = \Delta E/h$ that are also discrete in nature.

The treatment of molecules with more than two atoms is more complex; however, for future reference we mention that the vibrational normal modes of linear, symmetric triatomic molecules (such as carbon dioxide) and certain triatomic molecules that are not linear (such as water vapour and ozone) take fairly simple forms, as illustrated in Figure 3.8. These normal modes are designated by the labels v_1, v_2 and v_3, as shown†.

Wavelengths $hc/\Delta E_v$ associated with transitions between pure vibrational states are of order 1–20 μm, in the infra-red. Wavelengths $hc/\Delta E_J$ associated with transitions between pure rotational states would be of order 10^2–10^4 μm, in the far infra-red and microwave regions. In general, however, both vibrational and rotational energies change during a transition. The fact that the rotational energy differences are so much smaller than the vibrational energy differences means that, in terms of atmospheric spectra, rotational effects lead to 'fine-structure splitting' of vibrational spectral lines. Transitions from rotational sub-levels of one vibrational level to rotational sub-levels of another vibrational level therefore give rise to complex *vibration–rotation* spectral bands in the infra-red.

Wavelengths associated with transitions between electronically excited states are less than 1 μm and lie in the ultra-violet and

† The labels v_i do *not* refer to frequencies.

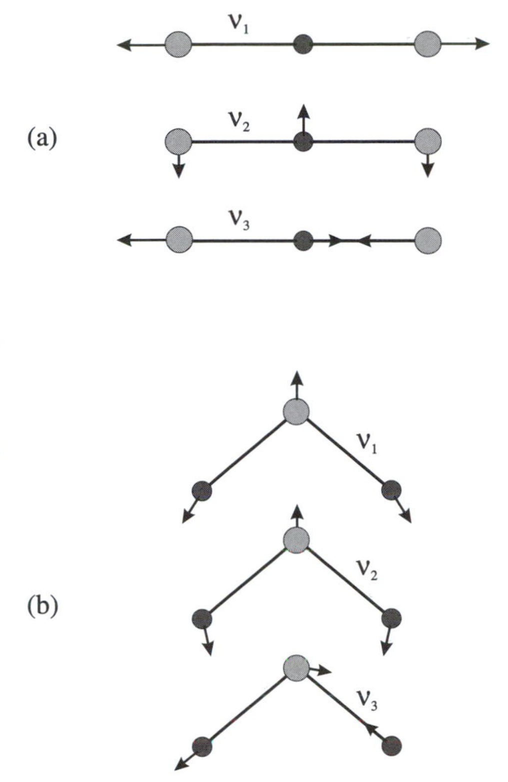

Figure 3.8 Schematic illustration of vibrational normal modes of triatomic molecules. (a) A linear triatomic molecule such as CO_2, showing the symmetric ν_1 and ν_2 modes and the asymmetric ν_3 mode (note that two independent forms of the ν_2 mode occur, one with displacements in the plane of the paper, as shown, and one with displacements perpendicular to the plane of the paper). (b) A non-linear triatomic molecule such as H_2O. Adapted after Herzberg (1945).

visible region. Thus, in the case of solar photons, we must also consider the electronically excited states of molecules. This again is a complex subject, but it can be described qualitatively in the case of a diatomic molecule, by using an energy diagram, such as Figure 3.9. The lower curved line shows the potential energy of the ground state versus the internuclear distance. It takes the form of a potential well resulting from electric repulsion at short distances and attraction at larger distances. The potential energy becomes constant at large enough distances, corresponding to dissociation of the molecule AB into two ground-state atoms, A and B. Within the potential well there is a number of vibrational energy levels, denoted by the equispaced horizontal lines. (Near the bottom of the well, the potential energy varies approximately quadratically with distance, so that the simple theory leading to (3.17) applies there.) A similar potential well applies to the first electronically excited state; however, in this case dissociation leads to one ground-state atom (A say) and one electronically excited atom (B^* say).

Several transitions are shown in Figure 3.9 by vertical arrows. These are (1) the absorption of a photon, giving rise to a transition from one vibrational level to another vibrational level in the ground-state molecule; (2) photo-dissociation to two ground-

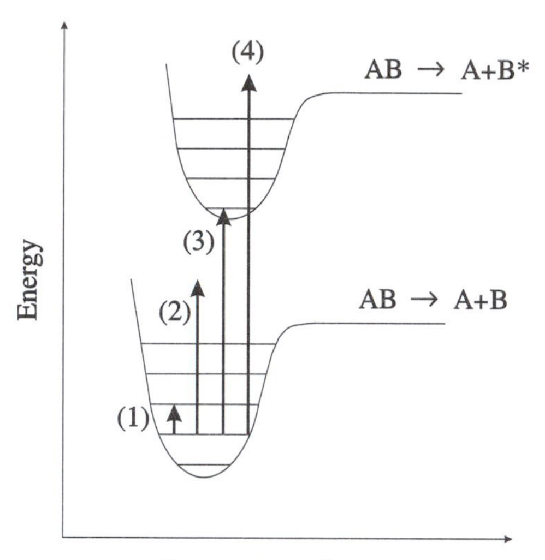

Figure 3.9 Schematic energy diagram for a diatomic molecule. See text for details.

state atoms; (3) absorption giving rise to a transition from a vibrational level of the ground state to a vibrational level in the excited state; and (4) photo-dissociation to one ground-state atom and one excited atom. Note that transitions (1) and (3) lead to discrete energy changes ΔE whereas (2) and (4) lead to continuous energy changes. It should be noted that transition (1) cannot occur for diatomic molecules with identical nuclei, such as N_2 and O_2, because such molecules have no permanent electric dipole moment.

3.3.2 Line shapes

In this section we again neglect scattering, so that only absorption contributes to the extinction coefficient k_ν, defined in Section 3.2.2. The implication of the simple pictures of vibrational and rotational transitions given in the previous section is that k_ν can generally be expected to vanish, except at certain discrete frequencies ν_n associated with transitions, for which k_ν is large; these frequencies correspond to *spectral lines*. This suggests a model

$$k_\nu = \sum_n S_n \delta(\nu - \nu_n), \tag{3.19}$$

where the S_n are constants, called the *line strengths*, and $\delta(\cdot)$ is the Dirac delta function.

However, this model of sharp, delta-function, peaks in the extinction coefficient is an idealisation. In practice several physical effects lead to a broadening of the sharp peaks (see Figure 3.10), so that

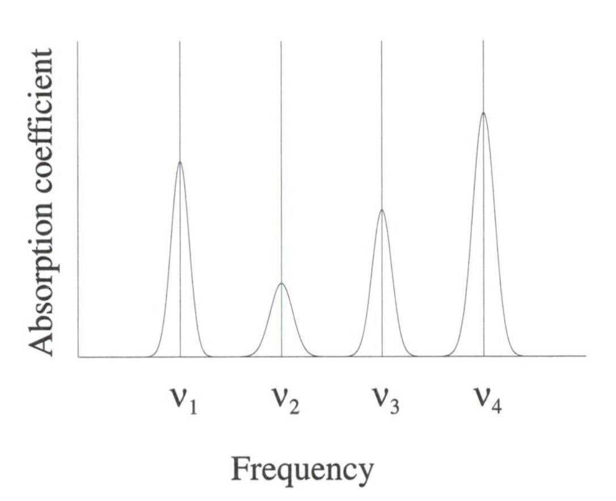

Figure 3.10 A schematic diagram of some broadened spectral lines centred at four frequencies, ν_1, ν_2, ν_3 and ν_4. Unbroadened lines would consist of sharp peaks at these frequencies, as indicated by the vertical lines.

the equation (3.19) is replaced by

$$k_\nu = \sum_n S_n f_n(\nu - \nu_n),$$

where the f_n are *line-shape functions*, each normalised such that

$$\int_{-\infty}^{\infty} f_n(\nu - \nu_n)\, d\nu = 1. \tag{3.20}$$

There are two main cases.

(a) *Collisional or natural broadening*. The energy levels have finite lifetimes for one of two reasons:

- there is a collisional lifetime τ_c (see Section 3.1.2);
- excited states have a natural lifetime τ_n with respect to spontaneous decay.

The finite lifetimes imply that the energy levels are not precisely defined, but rather have a spread of values of order $\hbar/\tau_c$ or $\hbar/\tau_n$, respectively, according to Heisenberg's Uncertainty Principle. There is thus a spread of values of each ΔE and hence of each transition frequency ν_n.

In the case of collisional broadening we obtain the *Lorentz line shape*, of the form

$$f(\nu - \nu_n) = \left(\frac{\gamma_L}{\pi}\right) \frac{1}{(\nu - \nu_n)^2 + \gamma_L^2}, \tag{3.21}$$

where $\gamma_L = (2\pi\tau_c)^{-1}$ is the half-width at half maximum of the line. Note that f satisfies equation (3.20). The derivation of equation (3.21) is quite complicated; the proof is omitted from this book, but is given in advanced texts on quantum mechanics or spectroscopy.

We can get τ_c and hence γ_L from kinetic theory, in terms of the mean molecular velocity (which is proportional to $T^{1/2}$), number density (proportional to p/T) and collision cross-section, giving the following temperature and pressure dependence of γ_L:

$$\gamma_L \propto p\, T^{-1/2}. \tag{3.22}$$

A similar calculation applies to natural broadening, but in this case $\gamma_L = (2\pi\tau_n)^{-1}$. In the infra-red, the corresponding line width is much less than that due to collisional broadening and than that due to the Doppler broadening discussed below, so natural broadening can usually be neglected. However, natural broadening can be important in the ultra-violet.

(b) *Doppler broadening.* The Doppler shift of a spectral line at a frequency $\nu = \nu_0$ due to a molecular velocity u of the emitter away from the observer is $\nu - \nu_0 = (u/c)\nu_0$, assuming that $u \ll c$. However the velocities follow the Maxwell–Boltzmann distribution; the probability that the speed of a molecule lies between u and $u+du$ is $P(u)\,du$, where

$$P(u) = \left(\frac{m}{2\pi kT}\right)^{1/2} \exp\left(-\frac{mu^2}{2kT}\right)$$

and m is the molecular mass. The spectral line shape, rather than being a delta function, becomes the convolution of $P(u)$ with a delta function:

$$\int_{-\infty}^{\infty} P(u)\delta\left(\nu - \nu_0 - \frac{u}{c}\nu_0\right) du \;\propto\; \exp\left(-\frac{mc^2(\nu - \nu_0)^2}{2kT\nu_0^2}\right).$$

We then get the *Doppler line shape*

$$k_\nu = \frac{S}{\gamma_D\sqrt{\pi}} \exp\left(-\frac{(\nu - \nu_0)^2}{\gamma_D^2}\right),$$

on normalising such that $S = \int_{-\infty}^{\infty} k_\nu\, d\nu$, where

$$\gamma_D = \frac{\nu_0}{c}\left(\frac{2kT}{m}\right)^{1/2}.$$

The half-width at half maximum is $\gamma_D(\ln 2)^{1/2}$, and is proportional to $T^{1/2}$ but independent of p.

The Lorentz and Doppler line shapes are shown in Figure 3.11; also shown is the Voigt line shape, which applies when both collisional and Doppler broadening are important. The fact that γ_L is proportional to pressure (see equation (3.22)) means that collisional broadening dominates at low altitudes (< 30 km), in the infra-red. Higher up, Doppler broadening becomes significant in the visible and ultra-violet.

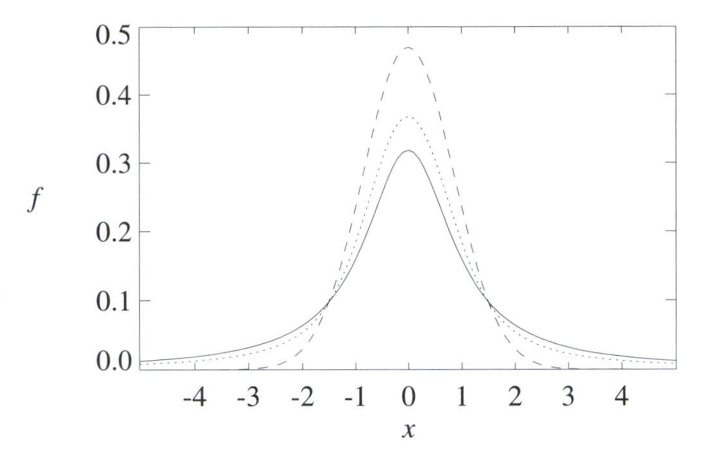

Figure 3.11 Illustrating the Lorentz (solid), Doppler (dashed) and Voigt (dotted) line shapes as a function of $x = (\nu - \nu_0)/\alpha$, where α is the half-width at half maximum appropriate for each shape. The curves are normalised such that the area under each is the same.

3.4 Transmittance

An important quantity that arises in the solution of the radiative transfer equation in Section 3.2.2 and the calculation of heating rates in Section 3.6 is the fraction of the spectral radiance leaving one point that arrives at another point. This fraction is represented by the *transmittance* or *transmission function*. For a parallel beam leaving point s_1 and arriving at point s_2 (or vice versa), the *spectral transmittance* is

$$\mathcal{T}_\nu(s_1, s_2) = \exp\left(-\left|\int_{s_1}^{s_2} k_\nu(s)\rho_a(s)\,ds\right|\right) = \exp[-|\chi_\nu(s_2) - \chi_\nu(s_1)|]; \tag{3.23}$$

cf. equations (3.11) and (3.13). The modulus signs express the fact that the fraction is independent of whether the radiation goes from s_1 to s_2 or from s_2 to s_1 and ensure that the fraction is ≤ 1. If scattering is neglected, the extinction coefficient k_ν depends on s through its temperature and pressure dependences. However, if the variation of k_ν along the path can be ignored, such as, for example, for measurements under laboratory conditions at fixed p and T, then equation (3.23) gives

$$\mathcal{T}_\nu(s_1, s_2) = \exp[-k_\nu(p, T)u_a(s_1, s_2)], \tag{3.24}$$

where

$$u_a(s_1, s_2) = \left|\int_{s_1}^{s_2} \rho_a(s)\,ds\right|.$$

Here u_a is the mass of absorber gas, per unit transverse cross-sectional area, in the path. If the absorber density is also constant along the path, then $u_a = \rho_a l$, where l is the length of the path.

Figure 3.12 shows examples of the transmittance $\mathcal{T}_\nu$ as a function of ν for three values of $q = u_a S/(\gamma_L \pi)$ for a Lorentz line at fixed p

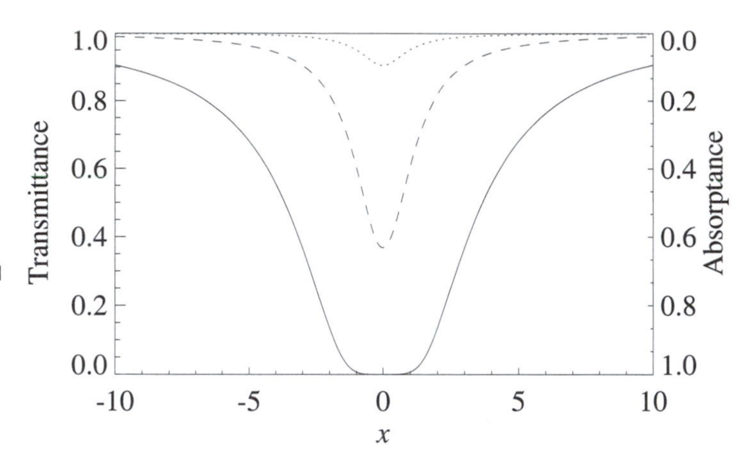

Figure 3.12 The transmittance $\mathcal{T}_\nu$ and absorptance $\mathcal{A}_\nu$ as functions of $x = (\nu - \nu_0)/\gamma_L$ for a Lorentz line, for three values of the quantity $q = u_a S/(\gamma_L \pi)$: $q = 10$, solid curve; $q = 1$, dashed curve; and $q = 0.1$, dotted curve.

and T. Also indicated is the *absorptance* $\mathcal{A}_\nu = 1 - \mathcal{T}_\nu$. Note that, for large amounts of absorber ($q \gg 1$, solid curve) the transmittance is effectively zero near the line centre, whereas for small amounts of absorber ($q \ll 1$, dotted curve) the transmittance is close to 1 for all frequencies.

It is often useful to average the transmittance over a spectral band, of width $\Delta\nu_r$, say, perhaps containing many spectral lines, to get the *band transmittance*

$$\overline{\mathcal{T}}_r = \frac{1}{\Delta\nu_r} \int_{\Delta\nu_r} \mathcal{T}_\nu \, d\nu. \tag{3.25}$$

Two other useful quantities are the *band absorptance*

$$\overline{\mathcal{A}}_r = 1 - \overline{\mathcal{T}}_r,$$

and the *equivalent width* or *integrated absorptance*

$$\mathcal{W}_r = \int_{\Delta\nu_r} (1 - \mathcal{T}_\nu) \, d\nu = \Delta\nu_r (1 - \overline{\mathcal{T}}_r) = \Delta\nu_r \overline{\mathcal{A}}_r. \tag{3.26}$$

If the integration is over a single broadened spectral line, then $\mathcal{W}_r$ represents the width of a rectangular line shape within which total absorption takes place, which has the same area as the actual line.

Note that, for small amounts of absorber (again at fixed p and T along the path) we have $\mathcal{T}_\nu = e^{-k_\nu u_a} \approx 1 - k_\nu u_a$, so that

$$\mathcal{W}_r \approx \int_{\Delta\nu_r} k_\nu u_a \, d\nu = S u_a, \tag{3.27}$$

using $S = \int_{\Delta\nu_r} k_\nu \, d\nu$. This is called the *weak-line approximation.* At the opposite extreme, for large amounts of absorber, the equivalent width can be calculated explicitly for a single Lorentz or Doppler line shape, assuming that k_ν is constant, to obtain the

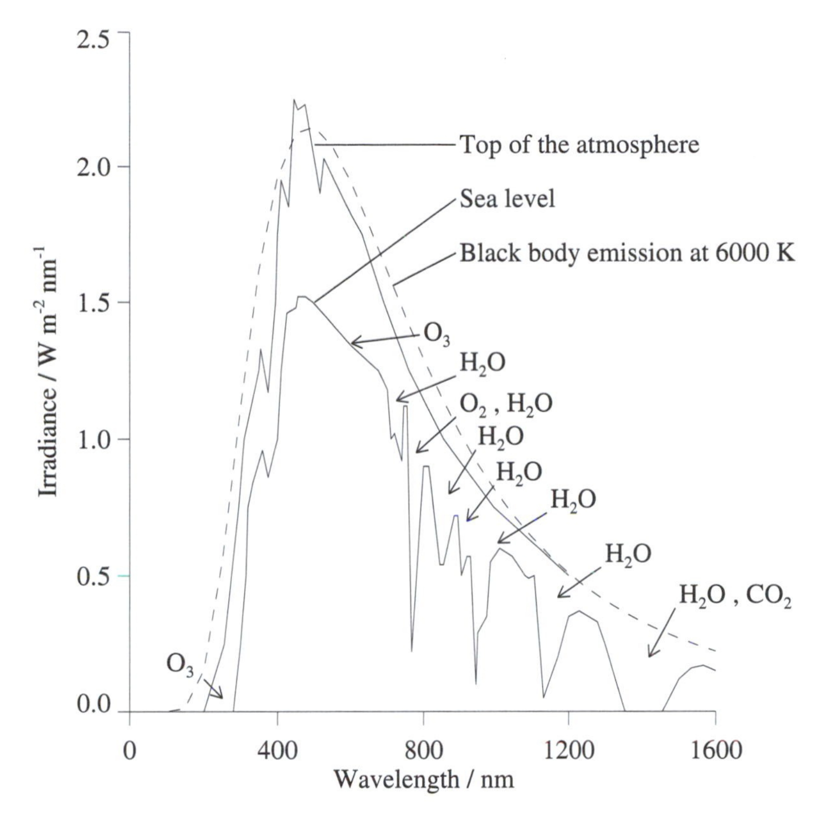

Figure 3.13 The irradiance spectrum of solar radiation at the top of the atmosphere and at sea level, compared with the black-body irradiance spectrum (dashed line), given by $B_\lambda(T)$ times the solid angle subtended by the Sun, for $T = 6000$ K. Adapted after Figure 4.6 of Brasseur and Solomon (1986).

strong-line approximations $W \approx 2(Su_a\gamma_L)^{1/2}$ for a Lorentz line and $W \approx 2\gamma_D\{\ln[Su_a/(\gamma_D\sqrt{\pi})]\}^{1/2}$ for a Doppler line; see Problems 3.6 and 3.8.

3.5 Absorption by atmospheric gases

3.5.1 The solar spectrum

Figure 3.13 shows the spectral irradiance of solar radiation at the top of the atmosphere (TOA) and at sea level, compared with the black-body spectral irradiance at a typical solar photospheric temperature of 6000 K. It is seen that the TOA irradiance is fairly close to the black-body irradiance at most wavelengths. However, absorption and scattering in the atmosphere cause significant deviations between the TOA irradiance and the sea-level irradiance, with especially large deviations (apparent in the sharp dips in the sea-level curve) at certain wavelengths, which are identified with particular absorbing gases, notably ozone (O_3) in the ultra-violet and visible and carbon dioxide (CO_2) and water vapour (H_2O) in the infra-red. We investigate atmospheric absorption in more detail in the next two sections.

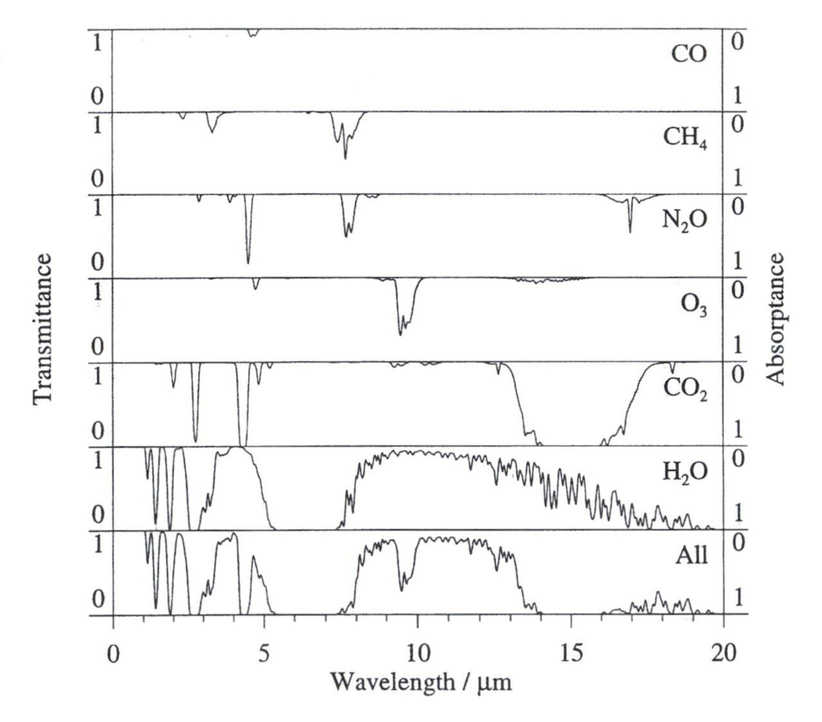

Figure 3.14 Infra-red absorption spectra for six strongly absorbing gases and for the six gases combined, for a vertical beam passing through the atmosphere, in the absence of clouds. Drawn from data supplied by Dr A. Dudhia.

3.5.2 Infra-red absorption

The absorption of infra-red radiation by the six most significant gaseous absorbers is conveniently summarised in Figure 3.14. This figure shows the transmittance for a vertical beam passing through the whole atmosphere, as a function of wavelength. The gases shown are all minor constituents (that is, they are present in much smaller quantities than molecular nitrogen (N_2) and molecular oxygen (O_2); see Section 2.2 and Table 2.1) and all but ozone (O_3) are concentrated mainly in the troposphere. The major constituents, N_2 and O_2, are not strongly radiatively active in the infra-red because they possess no permanent electric dipole moment, on account of their symmetry; see the end of Section 3.3.1. Since, on the scale of the diagram, much of the fine structure associated with individual spectral lines is not shown, the diagram can be regarded as plots of the band transmittance $\overline{\mathcal{T}}_r$ (equal to $1 - \overline{\mathcal{A}}_r$, where $\overline{\mathcal{A}}_r$ is the band absorptance) between the top of the atmosphere and the ground, corresponding to band widths $\Delta\nu_r$ associated with wavelength differences $\sim 0.1\,\mu$m.

The bottom panel of Figure 3.14 shows the total long-wave absorptance due to all gases. There is a broad region from 8 to 12 µm, called the *atmospheric window*, within which absorption is weak, except for a band near 9.6 µm associated with O_3.

Water vapour (H_2O) absorbs strongly over a wide band of wavelengths near 6.3 μm (associated with transitions involving the v_2 vibrational mode: see Figure 3.8) and over a narrower band near 2.7 μm (associated with the v_1 and v_3 vibrational modes). At longer wavelengths, especially beyond 16 μm, rotational transitions of H_2O become important, leading to strong absorption.

Carbon dioxide (CO_2) is a strong absorber in a broad band near 15 μm, associated with the vibrational v_2 'bending' mode, and in a narrower band near 4.3 μm, associated with the v_3 'asymmetric stretching' mode. (The band near 2.7 μm has a more complex origin.)

Ozone (O_3) absorbs strongly near 9.6 μm (associated with the v_1 and v_3 vibrational modes), in the atmospheric window. Since the other gases do not absorb significantly in this spectral region, ozone (which is mainly concentrated in the stratosphere) can therefore exchange radiation with the lower atmosphere; see Section 3.6.4.

Figure 3.14 gives information on the absorption of infra-red radiation over the total depth of the atmosphere, but not directly about the way in which the absorption varies with altitude. Moreover, it must be remembered that atmospheric gases also emit infra-red radiation and that this emission also varies with altitude. The vertical profiles of the absorption and emission are required in the calculation of the resulting heating and cooling; this is discussed in Section 3.6.3.

3.5.3 Ultra-violet absorption

In the ultra-violet, the main absorbers are molecular oxygen (O_2) and ozone. Absorption at these wavelengths is often depicted in terms of the *absorption cross-section* σ_v, which is equal to the absorption coefficient a_v times the molecular mass. Unlike in the infra-red, we must take account of electronic transitions, as well as vibrational and rotational transitions, when considering absorption at discrete wavelengths; moreover, photo-dissociation and photo-ionization (see Section 3.1) lead to important *continuum absorption*, i.e., absorption over a continuous range of wavelengths, rather than at discrete wavelengths.

The absorption cross-section for O_2 (Figure 3.15) has large values due to ionization at wavelengths below 100 nm; in the range 100–130 nm there are irregular bands of unknown origin. The Schumann–Runge continuum, in the range 130–175 nm, is due to the dissociation $O_2 \rightarrow O(^3P) + O(^1D)$, in which one oxygen atom remains in the ground 'triplet-P' state and the other goes to the excited 'singlet-D' state. The Schumann–Runge bands, in the range 175–200 nm, are associated with an electronic transition and superimposed vibrational transitions. The Herzberg continuum is found in the range 200–242 nm. At 242 nm dissociation into two ground-state oxygen atoms occurs; although this is an insignificant

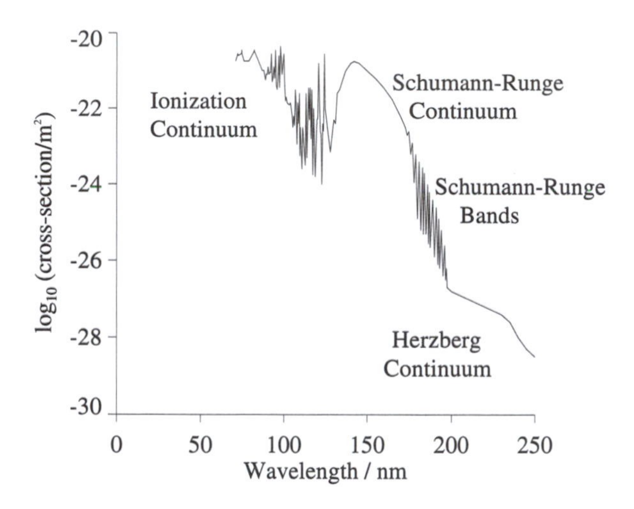

Figure 3.15 The absorption cross-section as a function of wavelength for O_2. Adapted after Figure 4.26 of Brasseur and Solomon (1986).

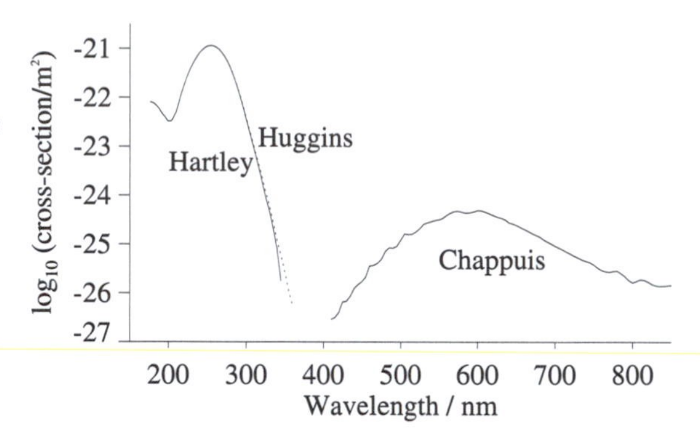

Figure 3.16 The absorption cross-section as a function of wavelength for O_3. Details of the fine structure of the Huggins band have been suppressed. In the Huggins band the solid line corresponds to a temperature of 203 K and the dashed line to a temperature of 273 K.

absorption feature, it is, however, very important in the formation of ozone (see Section 6.5.1). A further electronic transition gives rise to the weak Herzberg bands in the range 242–260 nm.

The ozone absorption cross-section (Figure 3.16) exhibits two continua in the ultra-violet and one in the visible and near infra-red, all due to photo-dissociation: the Hartley band, in the range 200–310 nm, the Huggins bands, in the range 310–350 nm and, in the visible and near infra-red, the Chappuis bands, in the range 400–850 nm. Although the absorption cross-section for the Chappuis bands is much smaller than those for the Hartley and Huggins bands, the Chappuis bands are important since they occur near the peak of the solar spectrum and absorb in the troposphere and lower stratosphere. Shorter-wavelength (more energetic) radiation is almost absent at these levels, since it is mostly absorbed higher up. This is demonstrated by Figure 3.17, which shows the altitude of unit optical depth (the peak of the Chapman layer in absorption; see Figure 3.18) as a function of wavelength.

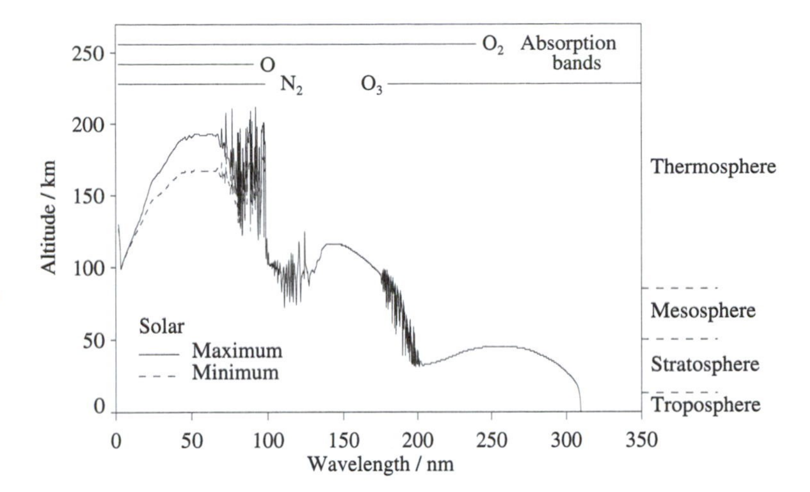

Figure 3.17 The altitude of unit optical depth for vertical solar radiation. The principal absorption bands are shown. Adapted after Meier (1991); earlier versions of this figure appeared in publications by Herzberg (1965), Chamberlain and Hunten (1987) and Lean (1987). Figure courtesy of Dr J. Lean and Dr R. Meier.

The heating of the atmosphere due to absorption of ultra-violet radiation is discussed in Section 3.6.2. In contrast to the infra-red, there is no significant emission from atmospheric gases in the ultra-violet, since the black-body spectral radiance at terrestrial temperatures is so small there; see Figures 3.1 and 3.2.

3.6 Heating rates

3.6.1 Basic ideas

One of the main goals of radiative calculations is to obtain radiative-heating rates throughout the atmosphere. For this one requires knowledge of the heating due to absorption of solar (short-wave) photons and the heating and cooling due to absorption and emission of thermal (long-wave) photons. In this section we consider some basic ideas; these are applied to solar and thermal radiation in later sections.

Consider a horizontal slab of atmosphere, of horizontal area A, at height z and of thickness Δz and make the plane-parallel atmosphere assumption. The upward power entering the bottom of the slab is $AF_z(z)$ and the upward power emerging at the top is $AF_z(z+\Delta z)$. The loss of radiative power within the volume $A\,\Delta z$ of the slab is therefore $A\,[F_z(z) - F_z(z+\Delta z)] \approx -(A\Delta z)\,dF_z/dz$. This loss of radiative power implies that radiative *diabatic heating* (see Sections 2.4 and 4.10) of the slab is occurring at a rate $-dF_z/dz$ per unit volume or

$$Q = -\frac{1}{\rho(z)}\frac{dF_z}{dz} \tag{3.28}$$

per unit mass, where ρ is the density of air. The units of Q are

$\mathrm{W\,kg^{-1}}$; often the quantity Q/c_p arises in calculations of the dynamical effects of radiative heating, where c_p is the specific heat capacity at constant pressure (cf. equation (4.35)), and this quantity has units $\mathrm{K\,s^{-1}}$. Since $F^{\uparrow}$ and $F^{\downarrow}$ both involve integration of spectral irradiances over frequency, F_z and Q also comprise contributions from different frequency bands.

3.6.2 Short-wave heating

Consider the diabatic heating rate per unit volume, ρQ_ν^{sw}, produced by absorption of short-wave solar radiation of frequency ν by a gas of density $\rho_{\mathrm{a}}(z)$ and extinction coefficient $k_\nu(z)$; scattering will be neglected. Assuming that the Sun is directly overhead, the appropriate optical path for each frequency is the *optical depth*, measured vertically downwards from the top of the atmosphere (taken to be $z=\infty$)†,

$$\chi_\nu(z) = \int_z^\infty k_\nu(z')\rho_{\mathrm{a}}(z')\,dz'. \tag{3.29}$$

By analogy with the second term of equation (3.13), the downward irradiance of the solar radiation is given by

$$F_\nu^{\downarrow}(z) = F_{\nu\infty}^{\downarrow} e^{-\chi_\nu(z)}, \tag{3.30}$$

where $F_{\nu\infty}^{\downarrow}$ is the downward solar irradiance at the top of the atmosphere. (Some care is needed to ensure that correct signs are obtained in equations (3.29) and (3.30)). Since scattering is neglected, the upward solar irradiance $F_\nu^{\uparrow}$ must be zero and the net vertical irradiance at frequency ν is

$$F_{z\nu}(z) = -F_{\nu\infty}^{\downarrow} e^{-\chi_\nu(z)}. \tag{3.31}$$

(Note incidentally that the exponential term $e^{-\chi_\nu(z)} = \mathcal{T}(z,\infty)$, the transmittance between height z and the top of the atmosphere.) The contribution to the heating rate per unit volume from this frequency is therefore

$$\rho Q_\nu^{\mathrm{sw}} = \frac{d}{dz}\left(F_{\nu\infty}^{\downarrow} e^{-\chi_\nu(z)}\right) = F_{\nu\infty}^{\downarrow}\left(-\frac{d\chi_\nu}{dz}\right)e^{-\chi_\nu(z)} = F_{\nu\infty}^{\downarrow} k_\nu(z)\rho_{\mathrm{a}}(z)e^{-\chi_\nu(z)}.$$

Suppose now that the extinction coefficient k_ν is independent of z and that the density of the absorber decays exponentially with height, $\rho_{\mathrm{a}}(z) = \rho_{\mathrm{a}}(0)e^{-z/H_{\mathrm{a}}}$, where H_{a} is constant. Then the integral in equation (3.29) can be evaluated explicitly, giving

$$\chi_\nu(z) = H_{\mathrm{a}} k_\nu \rho_{\mathrm{a}}(0) e^{-z/H_{\mathrm{a}}} = \chi_\nu(0)e^{-z/H_{\mathrm{a}}};$$

† Note that, for the direct solar beam, we do not integrate over solid angle, so that the diffuse approximation of Section 3.2.3 is not made.

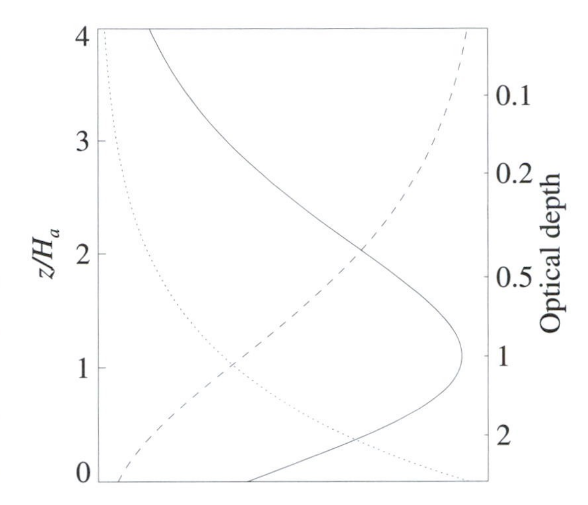

Figure 3.18 Vertical profiles of the short-wave volume heating rate, $\rho Q_\nu^{\mathrm{sw}}(z)$ (solid line), the negative of the vertical irradiance, $-F_{z\nu}(z)$ (dashed line) and the absorber density $\rho_{\mathrm{a}}(z)$ (dotted line), for solar radiation at frequency ν in the simple example described in the text. The horizontal scales are arbitrary. The left-hand ordinate shows the height z, divided by H_{a}; the right-hand ordinate shows the optical depth, $\chi_\nu(z)$. The optical depth of the ground $z = 0$ is arbitrarily chosen to be 3. The Sun is taken to be overhead. The absorber gas has a constant extinction coefficient and an exponentially decaying density $\rho_{\mathrm{a}} \propto \exp(-z/H_{\mathrm{a}})$. Note that, in this example, the optical depth has the same exponential variation as the absorber density.

this shows how the optical depth increases as the solar radiation penetrates downwards, i.e., as z decreases. Substitution into equation (3.31) then gives the vertical irradiance

$$F_{z\nu} = -F_{\nu\infty}^{\downarrow} \exp\left[-\chi_\nu(0)e^{-z/H_{\mathrm{a}}}\right]$$

and differentiation gives the monochromatic volume heating rate,

$$\rho Q_\nu^{\mathrm{sw}}(z) = F_{\nu\infty}^{\downarrow} k_\nu \rho_{\mathrm{a}}(0) \exp\left(-\frac{z}{H_{\mathrm{a}}} - \chi_\nu(0)e^{-z/H_{\mathrm{a}}}\right). \qquad (3.32)$$

Figure 3.18 shows the variation of the optical depth χ_ν, the vertical irradiance $F_{z\nu}$ and the volume heating rate ρQ_ν^{sw}, as functions of height in this simple example; note that the volume heating rate has a broad single peak. Differentiation of equation (3.32) with respect to z shows that ρQ_ν^{sw} has a maximum at the height where $\chi_\nu(z) = 1$, that is, the height where the optical depth equals unity. (We assume that the absorber is sufficiently 'optically thick' for this level to occur above the ground.)

A vertical structure of the type given by equation (3.32) is said to exhibit a *Chapman layer*. The peaked shape of the Chapman layer in the heating rate can be interpreted as follows. At high levels, above the peak, there is a large vertical irradiance (since little absorption of the solar beam has yet occurred), but few absorber molecules; at low levels, below the peak, there is a small vertical irradiance (since much absorption has occurred) but many absorber molecules. In each case the heating rate is small. However, near the level of unit optical depth both the irradiance and the absorber density are significant and so the heating rate is comparatively large. Chapman layers also occur in other processes determined primarily by the absorption of radiation; an example is the photo-dissociation that contributes to the formation of the 'ozone layer'; see Section 6.4.

3.6.3 Long-wave heating and cooling

We now consider the effects of thermal (long-wave) photons, allowing for both downward and upward paths. It can be shown, by solving the radiative transfer equation and integrating over solid angle, that the upward thermal irradiance at frequency ν and height z is

$$F_\nu^\uparrow(z) = \pi \int_0^z B_\nu(z') \frac{\partial \mathcal{T}_\nu^*(z',z)}{\partial z'}\, dz' + \pi B_\nu(0-)\mathcal{T}_\nu^*(0,z). \qquad (3.33)$$

Here $\mathcal{T}_\nu^*(z',z)$ is the spectral transmittance, averaged over the upward hemisphere to take account of all slanting paths between heights z' and z, and $B_\nu(0-)$ is the Planck function evaluated at the temperature of the Earth's surface. (Note that this temperature may be different from that of the air just above the surface.) The assumption of LTE has been made, so that the source function $J_\nu = B_\nu$, and the Earth's surface has been assumed to radiate as a black body. Similarly, the downward irradiance is

$$F_\nu^\downarrow(z) = -\pi \int_z^\infty B_\nu(z') \frac{\partial \mathcal{T}_\nu^*(z',z)}{\partial z'}\, dz';$$

there is no 'boundary' term here since the downward thermal irradiance at the top of the atmosphere is zero.

The net upward long-wave spectral irradiance is $F_{z\nu}(z) = F_\nu^\uparrow(z) - F_\nu^\downarrow(z)$ and from this the net long-wave diabatic heating rate per unit mass, Q_ν^{lw}, can be calculated using equation (3.28). The resulting expression for Q_ν^{lw} is quite complicated, but has a simple physical interpretation; namely the net heating or cooling at a given level is due to the difference between the energy *gained* per unit time by absorption of photons from neighbouring levels and the Earth's surface and the energy *lost* per unit time by emission of photons to neighbouring levels, the Earth's surface and space.

These heating and cooling terms can also be obtained directly. Consider for example the rate of loss of energy by a horizontal slab of atmosphere of thickness Δz and horizontal area A at height z by emission of photons to space – the *cooling-to-space* term. The derivation of the radiative transfer equation (3.10) shows that the spectral power emitted in a vertical direction from this slab is $k_\nu \rho_a J_\nu A\, \Delta z$, where J_ν is the source function, equal to B_ν under LTE, as above. The fraction of this power that escapes to space is given by the transmittance $\mathcal{T}_\nu(z,\infty) = \exp(-\int_z^\infty k_\nu \rho_a\, dz')$. Noting that

$$\frac{\partial \mathcal{T}_\nu(z,\infty)}{\partial z} = k_\nu(z)\rho_a(z)\mathcal{T}_\nu(z,\infty),$$

we find that the power escaping to space from the slab in a purely vertical beam is

$$B_\nu(z)\frac{\partial \mathcal{T}_\nu(z,\infty)}{\partial z} A\, \Delta z.$$

Now, integrating over all slanting paths as above and replacing $\mathcal{T}_\nu$ by $\mathcal{T}_\nu^*$, we obtain a contribution to the heating rate per unit mass

$$Q_\nu^{\text{cts}}(z) = -\frac{\pi B_\nu(z)}{\rho(z)} \frac{\partial \mathcal{T}_\nu^*(z, \infty)}{\partial z}. \tag{3.34}$$

The factor π in this equation comes from integration over the upward hemisphere and the minus sign from the fact that the power loss to space implies a negative heating at height z. The other contributions to the heating rate at height z can be calculated in similar, but more complicated, ways; these must include the heating of the slab due to absorption of photons emitted at other levels. A useful simplification for some purposes is the *cooling-to-space approximation*, in which the loss of photon energy to space dominates the other contributions; therefore, under this approximation, $Q_\nu^{\text{lw}} \approx Q_\nu^{\text{cts}}$.

All gases that absorb and emit at frequency ν must in principle be included in $\mathcal{T}_\nu^*$. Then $Q_\nu^{\text{lw}}(z)$ must be integrated over all relevant frequencies to obtain the total long-wave cooling $Q^{\text{lw}}(z)$.

3.6.4 Net radiative-heating rates

The short-wave and long-wave contributions to the diabatic heating rate Q can be computed using the principles described in the two previous sections, together with information on the atmospheric temperature structure and the concentration and spectroscopic characteristics of absorber gases. In this section we summarise the basic results of such computations.

Figure 3.19 shows the vertical profiles of the global-mean short-wave heating rate Q^{sw}/c_p and the long-wave cooling rate $-Q^{\text{lw}}/c_p$, in convenient units of K day^{-1}. The corresponding profiles of heating and cooling due to the most important atmospheric gases are also shown. It is clear that the total heating is approximately equal to the total cooling over much of the profile, except in the troposphere and above about 90 km. In the global mean, the middle atmosphere is therefore close to being in *radiative equilibrium*.

Below 80 km the short-wave heating rate is dominated by a Chapman-layer-like structure, centred near 50 km, due to absorption of solar radiation by ozone. The peak of the heating rate, at over 10 K day^{-1}, lies above the maxima both in the ozone density (near 22 km) and in the ozone mixing ratio (near 37 km). (The fact that this peak is above the maximum ozone density is consistent with Figure 3.18 and is explained qualitatively by the argument at the end of Section 3.6.2.) Below 15 km, in the troposphere, the main contribution to the heating rate is from water vapour, at about 1 K day^{-1}. Heating due to absorption by ozone and molecular oxygen is important between 80 and 100 km.

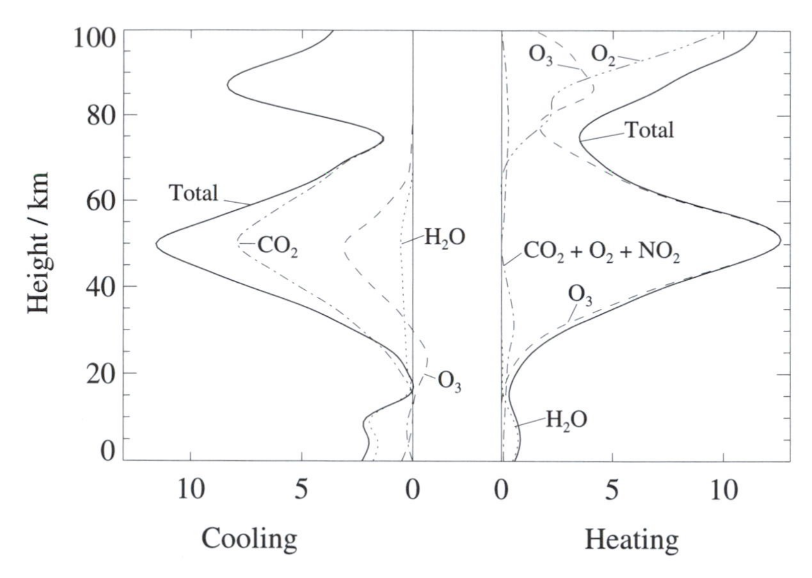

Figure 3.19 Global-mean vertical profiles of the short-wave heating rate and the long-wave cooling rate, in K day^{-1}, including contributions from individual gases. Adapted after London (1980).

The peak in long-wave cooling near 50 km has significant contributions due to carbon dioxide and ozone, both cooling to space. The dominant wavelength bands involved are 15 μm for CO_2 and 9.6 μm for O_3. A small amount of long-wave heating appears near 20 km, in the lower stratosphere, due to absorption by ozone of upwelling radiation from the troposphere at wavelengths near 9.6 μm, in the atmospheric window. Tropospheric cooling is dominated by water vapour, at about 2 K day^{-1}.

Figure 3.19 omits complications due to clouds and aerosols. Furthermore, the approximate radiative equilibrium in the global mean does not apply to individual latitudes and seasons. For example, in the winter stratosphere and the summer upper mesosphere there are big differences between Q^{sw} and $-Q^{\mathrm{lw}}$. In such regions dynamical heat transport is also significant; see Chapters 4 and 5.

It must be emphasised that the net heating rate $Q = Q^{\mathrm{sw}} + Q^{\mathrm{lw}}$ should *not* be thought of as a pre-ordained heating, to which the atmospheric temperature and wind fields respond. One reason is that Q itself depends strongly on the temperature, especially through Q^{lw}. A highly simplified, but nevertheless illuminating, model is to regard Q at a point $\boldsymbol{r}$ as a function of the local value of T and possibly of other variables. In radiative equilibrium $Q = 0$ by definition. Suppose that the corresponding radiative-equilibrium temperature field is $T_{\mathrm{r}}(\boldsymbol{r})$; then $Q(T_{\mathrm{r}}(\boldsymbol{r})) = 0$. If now the temperature deviates slightly from radiative equilibrium, $T = T_{\mathrm{r}} + \delta T$, say, the net heating rate will also differ from zero:

$$Q(T_{\mathrm{r}} + \delta T) \approx Q(T_{\mathrm{r}}) + \delta T \left.\frac{\partial Q}{\partial T}\right|_{T=T_{\mathrm{r}}} = \delta T \left.\frac{\partial Q}{\partial T}\right|_{T=T_{\mathrm{r}}} = -c_p \frac{\delta T}{\tau_{\mathrm{r}}} \tag{3.35}$$

say, since $Q(T_r) = 0$, where $\tau_r = -c_p(\partial Q/\partial T|_{T=T_r})^{-1}$. The net heating rate is therefore proportional and opposite in sign to the deviation of the actual temperature from the radiative-equilibrium temperature. Equation (3.35) is one form of the *Newtonian-cooling approximation* and the coefficient τ_r (which is positive in practice) is the *radiative relaxation time*. In this simple model (and in more realistic models) there is a kind of 'radiative spring', which tries to pull the temperature towards radiative equilibrium. This spring is opposed by other physical processes, especially dynamical processes, which often drive the atmosphere away from radiative equilibrium and thus force the net heating to be non-zero. In this sense we can regard the dynamics as driving the net heating, rather than the other way round; see also Section 8.5.2.

3.7 The greenhouse effect revisited

In Sections 1.3.2 and 1.3.3 we introduced some simple ideas concerning the *greenhouse effect*, the process by which the radiative properties of the atmosphere raise the surface temperature above that which would occur in the absence of a radiation-absorbing atmosphere (see Section 1.3.1). We considered there a shallow, highly idealised atmosphere.

We now consider a more complex model, involving a deep atmosphere. For simplicity, we assume that the atmosphere is transparent to solar radiation. To calculate the radiative transfer in the infrared, we use the diffuse approximation of Section 3.2.3 and further assume that the atmosphere is *grey*: that is, the extinction coefficient k (and hence the scaled optical depth χ^* (defined by equations (3.29) and (3.16))) is independent of frequency. We can therefore integrate equation (3.15) and the corresponding equation for upward irradiance over frequency to obtain the following equations for the spectrally integrated long-wave irradiances $F^{\uparrow}(\chi^*)$ and $F^{\downarrow}(\chi^*)$:

$$-\frac{dF^{\uparrow}}{d\chi^*} + F^{\uparrow} = \pi B, \tag{3.36a}$$

$$\frac{dF^{\downarrow}}{d\chi^*} + F^{\downarrow} = \pi B. \tag{3.36b}$$

Equations (3.36) represent the *two-stream approximation*; note that B is the spectrally integrated Planck function, so that $\pi B(T) = \sigma T^4$ by equation (3.7), where T is the temperature at the level corresponding to the scaled optical depth χ^*. In each of equations (3.36) the first term represents the rate of change of the irradiance along the path[†], the second term represents extinction and the term πB on

[†] The minus sign in equation (3.36a) appears because χ^* is an optical *depth* and *decreases* along the upward path.

the right-hand side represents emission, which contributes equally in both directions; cf. Section 3.2.2.

Given that the atmosphere is transparent to short-wave radiation, there can be no short-wave heating ($Q^{sw} = 0$); if we assume that the atmosphere is in *radiative equilibrium*, then the long-wave heating Q^{lw} is also zero; see Section 3.6.4. Hence, by equation (3.28), the net upward long-wave irradiance F_z is independent of height, so

$$F_z = F^{\uparrow} - F^{\downarrow} = \text{constant}.$$

The constant can be found by considering the boundary condition at the top of the atmosphere. Here the scaled optical depth $\chi^* = 0$ and the downward long-wave irradiance $F^{\downarrow}(0) = 0$ also, since the only incoming radiation is short-wave. The net upward long-wave irradiance here is therefore $F_z = F^{\uparrow}(0)$, which must balance the incoming unreflected short-wave irradiance, which we take as $F_0 \approx 240\,\text{W m}^{-2}$, as in Section 1.3.2. Hence

$$F_z = F^{\uparrow} - F^{\downarrow} = F_0 \qquad (3.37)$$

We now find $F^{\uparrow}$, $F^{\downarrow}$ and πB by a series of tricks. We first add equations (3.36a) and (3.36b) to get

$$-\frac{d}{d\chi^*}\left(F^{\uparrow} - F^{\downarrow}\right) + F^{\uparrow} + F^{\downarrow} = 2\pi B(T);$$

but, by equation (3.37), the derivative vanishes, so

$$\pi B(T) = \tfrac{1}{2}\left(F^{\uparrow} + F^{\downarrow}\right). \qquad (3.38)$$

We next subtract equation (3.36b) from equation (3.36a) and use equation (3.37) to get

$$\frac{d}{d\chi^*}\left(F^{\uparrow} + F^{\downarrow}\right) = F^{\uparrow} - F^{\downarrow} = F_0.$$

Since F_0 is constant, we can integrate to get

$$F^{\uparrow} + F^{\downarrow} = F_0\chi^* + \text{constant}.$$

Again using the upper boundary condition, we see that the constant is F_0, so that

$$F^{\uparrow} + F^{\downarrow} = F_0(1 + \chi^*). \qquad (3.39)$$

From equations (3.37) and (3.39) we can then find the upward and downward irradiances in terms of F_0 and the scaled optical depth,

$$F^{\uparrow} = \tfrac{1}{2}F_0(2 + \chi^*), \qquad (3.40)$$

$$F^{\downarrow} = \tfrac{1}{2}F_0(\chi^*),$$

and hence, using equation (3.38), an expression for the temperature:

$$\pi B(T) = \sigma T^4 = \tfrac{1}{2}F_0(1 + \chi^*). \qquad (3.41)$$

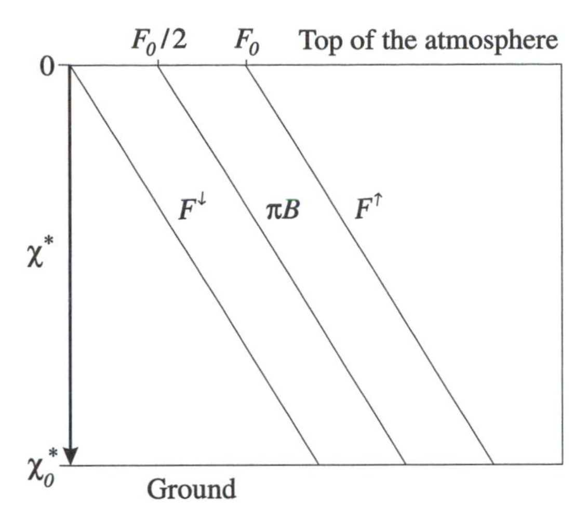

Figure 3.20 Results from a simple two-stream model. The sloping lines show the variations with scaled optical depth χ^* of the upward ($F^{\uparrow}$) and downward ($F^{\downarrow}$) spectrally integrated long-wave irradiances and of the spectrally integrated black-body irradiance πB.

Thus $F^{\uparrow}$, $F^{\downarrow}$ and B are all linear in the scaled optical depth; they are plotted in Figure 3.20.

The expressions derived above apply within the atmosphere. We now consider the radiation balance at the ground, which is taken to be a black body at scaled optical depth χ_0^* and with temperature T_{g}. From equation (3.40) the upward long-wave irradiance in the atmosphere just above the ground is $F_0(2 + \chi_0^*)/2$. If we assume that this irradiance is equal to that from the ground itself, the black-body value $\pi B(T_{\mathrm{g}}) = \sigma T_{\mathrm{g}}^4$, then

$$\pi B(T_{\mathrm{g}}) = \sigma T_{\mathrm{g}}^4 = F_0(1 + \tfrac{1}{2}\chi_0^*). \tag{3.42}$$

Note that this again demonstrates a greenhouse effect, since σT_{g}^4 is larger than F_0 (because the atmosphere has a non-zero optical depth, $\chi_0^* > 0$) and so T_{g} is larger than the value of 255 K that would apply in the absence of an absorbing atmosphere.

A somewhat unexpected aspect of this model is that there is a temperature discontinuity at the ground. If we take the temperature of the air just above the ground to be T_{b}, the equation (3.41) gives

$$\pi B(T_{\mathrm{b}}) = \sigma T_{\mathrm{b}}^4 = \tfrac{1}{2}F_0(1 + \chi_0^*). \tag{3.43}$$

Then equations (3.42) and (3.43) imply that

$$F_0 = 2\pi[B(T_{\mathrm{g}}) - B(T_{\mathrm{b}})] = 2\sigma(T_{\mathrm{g}}^4 - T_{\mathrm{b}}^4);$$

hence $T_{\mathrm{g}} > T_{\mathrm{b}}$, since $F_0 > 0$, so the ground temperature is larger than the temperature of the air at the bottom of the atmosphere. This, of course, is an unrealistic feature of the radiative model: inclusion of other physical effects, such as convection, would lead to a smoothing-out of the discontinuity.

3.8 A simple model of scattering

We conclude this chapter with a brief discussion of the scattering of solar radiation in the atmosphere. We consider a plane-parallel atmosphere and make the two-stream approximation, assuming that a fraction f of the spectral irradiance is scattered forwards (i.e., in the direction of the beam) and a fraction $1-f$ is scattered backwards. These fractions represent integrals over the forward- and backward-pointing hemispheres of the corresponding radiances.

We first work in terms of the vertical coordinate z and consider the downward penetration of the downward spectral irradiance $F_\nu^\downarrow$ at a short-wave frequency ν. By analogy with the radiative-transfer equation (3.10), this satisfies the equation

$$\frac{dF_\nu^\downarrow}{d(-z)} = -a_\nu \rho_a F_\nu^\downarrow - s_\nu \rho_a (1-f) F_\nu^\downarrow + s_\nu \rho_a (1-f) F_\nu^\uparrow. \tag{3.44}$$

This states that the downward irradiance decreases in the downward direction because of the absorption term $a_\nu \rho_a F_\nu^\downarrow$ and because of the backscattered component $s_\nu \rho_a (1-f) F_\nu^\downarrow$, and increases because of the backscattering of upward irradiance $s_\nu \rho_a (1-f) F_\nu^\uparrow$ into the downward direction. Emission is neglected, since the Planck function for short-wave frequencies at terrestrial atmospheric temperatures is negligible. A rearrangement of equation (3.44), in terms of the extinction coefficient $k_\nu = a_\nu + s_\nu$ (see equation (3.9)) gives

$$\frac{dF_\nu^\downarrow}{d(-z)} = -k_\nu \rho_a F_\nu^\downarrow + s_\nu \rho_a f F_\nu^\downarrow + s_\nu \rho_a (1-f) F_\nu^\uparrow.$$

We now replace $-k_\nu \rho_a \, dz$ by the optical depth increment $d\chi_\nu^*$; cf. equation (3.29). Introducing the *albedo for single scattering*,

$$\omega_\nu = \frac{s_\nu}{a_\nu + s_\nu} = \frac{s_\nu}{k_\nu},$$

and dropping the subscripts ν for simplicity, we obtain

$$\frac{dF^\downarrow}{d\chi^*} + F^\downarrow = \omega[fF^\downarrow + (1-f)F^\uparrow]. \tag{3.45a}$$

In the same way we obtain the equation for the upward irradiance $F^\uparrow$:

$$-\frac{dF^\uparrow}{d\chi^*} + F^\uparrow = \omega[fF^\uparrow + (1-f)F^\downarrow]. \tag{3.45b}$$

The simplest case is that of an atmospheric layer in which there is scattering, but no absorption, i.e., with $\omega = 1$. This layer could for example be a crude representation of a cloud. Equations (3.45)

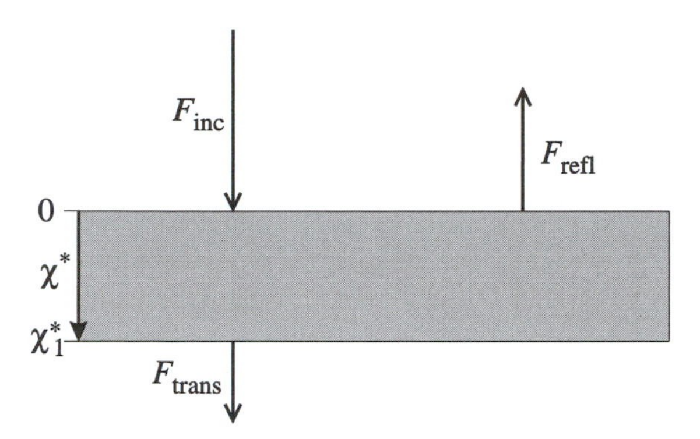

Figure 3.21 Scattering from a horizontal layer, showing the incoming spectral irradiance F_{inc}, the transmitted spectral irradiance F_{trans} and the reflected spectral irradiance F_{refl}.

then give

$$\frac{dF^{\downarrow}}{d\chi^*} + (1-f)\left(F^{\downarrow} - F^{\uparrow}\right) = 0, \tag{3.46a}$$

$$-\frac{dF^{\uparrow}}{d\chi^*} + (1-f)\left(F^{\uparrow} - F^{\downarrow}\right) = 0. \tag{3.46b}$$

Addition of equations (3.46) shows that the net downward spectral irradiance is constant:

$$-F_z = F^{\downarrow} - F^{\uparrow} = \text{ constant.} \tag{3.47}$$

Suppose that the downward spectral irradiance at frequency ν at the top of the layer ($\chi^* = 0$, say) is F_{inc} and that the upward spectral irradiance at the bottom of the layer ($\chi^* = \chi_1^*$, say) is zero. Suppose also that the transmitted spectral irradiance at the bottom is F_{trans} and the reflected spectral irradiance at the top is F_{refl}; see Figure 3.21. We therefore have the boundary conditions

$$F^{\downarrow}(0) = F_{inc}, \qquad F^{\uparrow}(0) = F_{refl}, \qquad F^{\downarrow}(\chi_1^*) = F_{trans}, \qquad F^{\uparrow}(\chi_1^*) = 0. \tag{3.48}$$

We can define a transmission coefficient $\mathsf{T} = F_{trans}/F_{inc}$ and a reflection coefficient (or layer albedo) $\mathsf{R} = F_{refl}/F_{inc}$. Then $-F_z = F_{inc} - F_{refl} = (1-\mathsf{R})F_{inc} = F_{trans} - 0 = \mathsf{T}F_{inc}$. Hence

$$\mathsf{R} + \mathsf{T} = 1, \tag{3.49}$$

implying conservation of irradiance.

Using $F^{\downarrow} - F^{\uparrow} = -F_z = \mathsf{T}F_{inc}$ and the upper boundary condition $F^{\downarrow}(0) = F_{inc}$, we can now solve equation (3.46a) to get

$$F^{\downarrow}(\chi^*) = F_{inc}[1 - (1-f)\mathsf{T}\chi^*] \tag{3.50a}$$

and, using equation (3.47),

$$F^{\uparrow}(\chi^*) = F_{inc}[\mathsf{R} - (1-f)\mathsf{T}\chi^*]. \tag{3.50b}$$

Then substitution of $F^{\uparrow}(\chi_1^*) = 0$ into equation (3.50b) and use of equation (3.49) gives the transmission and reflection coefficients

$$\mathsf{T} = \frac{1}{1+(1-f)\chi_1^*}, \qquad \mathsf{R} = \frac{(1-f)\chi_1^*}{1+(1-f)\chi_1^*}.$$

Note that, as $\chi_1^* \to \infty$, the reflection coefficient (layer albedo) tends to 1 and the transmission coefficient tends to 0, i.e., the layer becomes a nearly perfect reflector as the optical thickness becomes large.

The case of a layer that both absorbs and scatters solar radiation, so that the single-scattering albedo $\omega < 1$, involves more complicated mathematics, but the physical conclusions are similar, except that $\mathsf{R} + \mathsf{T} < 1$. See Problem 3.12.

References

For general introductions to the subject of atmospheric radiation see for example Houghton (1986), Wallace and Hobbs (1977), Salby (1996) and Goody (1995). The work by Goody and Yung (1989) is the second edition of the classic text in the field, and goes far beyond the treatment given in this chapter. The proof that the spectral energy density in an isothermal cavity depends only on the frequency and temperature is given in thermodynamics books, such as that by Adkins (1983). The derivations of the Planck function, equation (3.1), and the Boltzmann distribution, equation (3.4), are given in statistical mechanics books, such as those by Mandl (1988) and Guenault (1995). For the Poynting vector see standard books on electromagnetism, such as those by Grant and Phillips (1990) and Lorrain *et al.* (1988). Banwell and McCash (1994) and Atkins (1994) provide introductions to molecular spectroscopy, while an excellent treatment of many aspects of radiative transfer and spectroscopy is presented by Thorne (1988). Derivations of the Lorentz profile, equation (3.21), are given for example by Thorne (1988) and Goody and Yung (1989); see also Bransden and Joachain (1989) for the quantum-mechanical details. The quantum harmonic oscillator is covered for example by Rae (1992) and McMurry (1993). The cooling-to-space approximation was introduced by Rodgers and Walshaw (1966).

Problems

3.1. (i) Given equation (3.1) for B_ν, derive equation (3.2) for B_λ.

(ii) Show that the values of ν and λ that maximise $B_\nu(T)$ and $B_\lambda(T)$ are given by

$$\frac{\nu_{\max}}{T} = c_1, \qquad \lambda_{\max} T = c_2,$$

respectively, where c_1 and c_2 are constants. (Note that ν_{max} and λ_{max} do not correspond to the same photon energy.) Given that $c_2 = 2.9 \times 10^{-3}$ m K, find the temperatures for which B_λ is maximum at 500 nm and at 10 μm.

(iii) Assuming that the Sun behaves as a black body at a temperature of $T = 5800$ K, calculate $B_\nu(T)$ and $B_\lambda(T)$ at 500 nm and the percentage increase in these spectral radiances if the temperature increases by 100 K.

3.2. (i) Elements of surface ΔS_1 and ΔS_2 are a distance r apart, with normals inclined at angles θ_1 and θ_2 to the line joining them. Under what conditions is the net radiative power flow between them in the optical passband $\Delta\nu$ given by

$$\Delta P_{1\to 2} = \frac{B_\nu(T_1)}{r^2}\,\Delta\nu\,\Delta S_1\,\Delta S_2 \cos\theta_1 \cos\theta_2?$$

(ii) Integrate $\int_0^\infty B_\nu\,d\nu$ using the substitution $x = h\nu/(kT)$ and hence derive the right-hand member of equation (3.7). Use the relations

$$\int_0^\infty \frac{x^3\,dx}{e^x - 1} = \frac{\pi^4}{15}, \qquad \sigma = \frac{2\pi^5 k^4}{15h^3c^2}.$$

Hence use the results of (i) to show that the solar constant F_s (i.e., the normal irradiance just outside the atmosphere) equals $\sigma T_s^4 r_s^2/d^2$, where T_s is the effective black-body temperature of the Sun, r_s is the Sun's radius and d is the Sun–Earth distance. Evaluate F_s, given that $T_s = 5800$ K.

(iii) By carrying out the appropriate integration, find the power absorbed by a horizontal black disc of radius a from an infinite horizontal black plane at temperature T_1 over which it is suspended at height h.

(iv) Find the answer to (iii) using symmetry and thermodynamic arguments. Find also the power absorbed from the plane by a black sphere of radius a.

3.3. Assume that the Earth and Sun radiate as black bodies at 288 and 6000 K, respectively. Show that the radiative power per unit area, per unit wavelength interval (the spectral irradiance), falling on a horizontal plane just above the Earth's surface and with the Sun overhead, is $\pi B_\lambda(288\,\text{K})$ from the side facing the Earth and

$$(1 - A)\left(\frac{r_s}{d}\right)^2 \pi B_\lambda(6000\,\text{K})$$

from the side facing the Sun. Here $A \approx 0.3$ is the albedo of the Earth–atmosphere system (see Section 1.3.1), r_s is the Sun's radius and d is the Sun–Earth distance. Use Figure 3.1 to estimate the maximum values of the solar and terrestrial spectral irradiances and hence sketch the variations of these irradiances with wavelength.

3.4. A low-altitude Earth satellite in an equatorial orbit carries below it two isolated spherical radiation sensors of negligible heat capacity. One is painted white and may be assumed to be perfectly reflecting at all wavelengths at which there is significant solar energy ($\lambda < 4\,\mu\text{m}$) and perfectly absorbing at longer wavelengths. The other is painted black and is perfectly absorbing at all wavelengths. Assume that there is no input of direct or scattered energy from the spacecraft.

(i) Calculate the temperatures of the two spheres at midnight over a thick cloud sheet at a temperature of 280 K.

(ii) Find the radiance due to diffusely scattered sunlight just above a cloud of albedo α when the Sun is overhead.

(iii) Assuming that the spheres are shadowed from overhead direct sunlight by the spacecraft, calculate their temperatures over a thick cloud of temperature 270 K and albedo 0.8, given that the solar constant $F_s = 1370\,\text{W m}^{-2}$.

3.5. Use the two-stream approximation (Section 3.7) to investigate an atmosphere that is transparent to solar radiation, with the ground at temperature T_g. The only absorption within the atmosphere takes place in the infra-red; scattering may be ignored. The atmosphere contains a single absorbing gas, of constant mass mixing ratio μ and constant absorption coefficient k. Assume that the atmosphere is in radiative equilibrium and that the spectrally integrated irradiances $F^{\uparrow}$ and $F^{\downarrow}$ are continuous between the ground and the atmosphere just above the ground.

(i) Find an expression for χ^* as a function of the pressure p at level z.

(ii) Show that the spectrally integrated Planck function $B(T)$ is related to the spectrally integrated net upward irradiance F_z by

$$2\pi B(T) = \left(\frac{1.66k\mu p}{g(1+\mu)} + 1\right)F_z,$$

where g is the acceleration due to gravity.

(iii) Show that $F_z = 2\pi[B(T_g) - B(T_b)]$, where T_b is the temperature at the bottom of the atmosphere. (Note that, in this model, $T_b \neq T_g$ in general.)

(iv) Find F_z, given that at 250 hPa the temperature is 220 K and that $T_g = 280$ K. (Hint: you will need $\chi^*(250\,\text{hPa})/\chi^*(1000\,\text{hPa})$.)

(v) Find the temperature discontinuity $T_g - T_b$ at the surface. What does your answer imply about radiative equilibrium?

(vi) Plot $F^{\uparrow}$, $F^{\downarrow}$ and πB as functions of p.

3.6. Investigate the equivalent width of a Lorentz line, as follows. Substitute S times the right-hand side of equation (3.21) into equation (3.25) and use equation (3.26) to get the following expression for the equivalent width:

$$\mathcal{W} = \gamma_{\mathrm{L}} \int_{-\infty}^{\infty} \left[1 - \exp\left(-\frac{q}{x^2+1}\right)\right] dx \qquad (3.51)$$

where $x = (\nu - \nu_0)/\gamma_{\mathrm{L}}$ and $q = u_{\mathrm{a}} S/(\gamma_{\mathrm{L}} \pi)$.

The integral in equation (3.51) can be evaluated in terms of Bessel functions; however, there are two special cases in which it can be approximated easily. The first is the case of a *strong line*, which is valid when the mass of absorber in the path is so large that $u_{\mathrm{a}} S/(\gamma_{\mathrm{L}} \pi) \gg 1$, i.e., $q \gg 1$: sketch the integrand in equation (3.51) in this case. Verify that the exponential in the integrand is then negligible except where $x^2 \gtrsim q \gg 1$ and show that the equivalent width becomes

$$\mathcal{W} \approx 2\gamma_{\mathrm{L}} \int_{0}^{\infty} [1 - \exp(-q/x^2)] dx.$$

Use the substitution $y = 1/x$ and integration by parts to show that

$$\mathcal{W} \approx 4\gamma_{\mathrm{L}} q \int_{0}^{\infty} \exp(-qy^2)\, dy = 2\gamma_{\mathrm{L}}(\pi q)^{1/2} = 2(S u_{\mathrm{a}} \gamma_{\mathrm{L}})^{1/2}.$$

On the other hand, if the mass of absorber in the path is small, we have the *weak-line* limit, with $q \ll 1$. Show that, in this case,

$$\mathcal{W} \approx \gamma_{\mathrm{L}} \int_{-\infty}^{\infty} \frac{q\, dx}{x^2+1} = \gamma_{\mathrm{L}} q \pi = S u_{\mathrm{a}},$$

in agreement with equation (3.27).

3.7. Under what conditions on the absorber mixing ratio is the transmittance at the centre ν_0 of a Lorentz line, for a path at pressure p, of length l and absorber density ρ_{a}, independent of the pressure?

3.8. Investigate the equivalent width of a Doppler line, as follows. Put $r = u_{\mathrm{a}} S/\gamma_{\mathrm{D}} \sqrt{\pi}$ and $x = (\nu - \nu_0)/\gamma_{\mathrm{D}}$, to get

$$\mathcal{W} = \gamma_{\mathrm{D}} \int_{-\infty}^{\infty} \left[1 - \exp\left(-re^{-x^2}\right)\right] dx.$$

The integral cannot be evaluated explicitly in general. However, in the *strong-line* limit, $r \gg 1$, it can be shown that the absorptance $\mathcal{A}_\nu = 1 - \mathcal{T}_\nu$ (the integrand [...] above) is nearly zero for $|x| > (\ln r)^{1/2}$ and nearly unity for $|x| < (\ln r)^{1/2}$. Hence show that the integral can be evaluated approximately as

$$\mathcal{W} = 2\gamma_{\mathrm{D}} \int_{0}^{\infty} (1 - \mathcal{T}_\nu)\, dx \approx 2\gamma_{\mathrm{D}} (\ln r)^{1/2} = 2\gamma_{\mathrm{D}} \left[\ln\left(\frac{u_{\mathrm{a}} S}{\gamma_{\mathrm{D}} \sqrt{\pi}}\right)\right]^{1/2}.$$

Show that, in the *weak-line* limit $r \ll 1$,

$$\mathcal{W} \approx \gamma_D \int_{-\infty}^{\infty} r \exp(-x^2)\, dx = \gamma_D r \sqrt{\pi} = S u_a,$$

in agreement with equation (3.27).

3.9. Consider a CO_2 line at $15\,\mu m$ (wavenumber $\tilde{\nu} = 1/\lambda = 667\,cm^{-1}$) with $\gamma_L = (p/p_0)(T_0/T)^{1/2}\gamma_{L0}$ (cf. equation (3.22)), where $p_0 = 1000\,hPa$, $T_0 = 250\,K$ and $\gamma_{L0} = 3 \times 10^9\,Hz$. Find the approximate pressure level in the atmosphere at which the transmittance of a horizontal path at the line centre under the assumption of pure Doppler broadening is the same as that under the assumption of pure Lorentz broadening. (Assume that the atmosphere is at a typical temperature of 250 K.)

3.10. The Curtis–Godson approximation is a method of calculating approximate transmittances for paths of varying pressure and absorber density. For a vertical path, one defines a mean pressure $\overline{p}$ and mean temperature $\overline{T}$, weighted by the absorber density:

$$\overline{p} = \frac{1}{u_a} \int p \rho_a\, dz, \qquad \overline{T} = \frac{1}{u_a} \int T \rho_a\, dz,$$

where $u_a = \int \rho_a\, dz$ is the total mass per unit transverse cross-sectional area of the path.

Show that the spectral means of atmospheric transmittances calculated using the Curtis–Godson approximation are accurate when (i) strong pressure-broadening conditions exist or (ii) the absorption is weak, irrespective of the line shape.

Derive $\overline{p}$ for an atmospheric absorber with a constant mass mixing ratio μ for a vertical path from height z to ∞.

3.11. (i) Write down an expression for the heating rate h ($= Q/c_p$ in units of $K\,s^{-1}$) at height z in the atmosphere due to an absorber of density $\rho_a(z)$ and constant extinction coefficient k when the Sun is overhead. (Denote the normal solar spectral irradiance integrated over the absorption band by $F_{\nu s}\,\Delta\nu$; neglect scattering.)

(ii) A simple model of radiative heating by solar radiation in the upper stratosphere and lower mesosphere assumes that ozone has a mass mixing ratio

$$\mu(p) = a p^{1/2},$$

where a is a constant and p is pressure. Show that the heating rate h due to ozone is proportional to

$$p^{1/2} \exp\left(-\frac{2ak}{3g} p^{3/2}\right),$$

where k is the extinction coefficient of ozone for solar radiation. Show that the pressure p_m at which h is a maximum is close to that of the stratopause and evaluate $h(p_m)$ in $K\,day^{-1}$.

(Take $k = 1.5 \times 10^4$ m^2 kg^{-1} in a spectral region where the solar irradiance is 7 W m^{-2} and take $a = 3 \times 10^{-7}$ Pa$^{-1/2}$.)

(iii) In a simple model of radiative cooling due to thermal emission by CO_2 ('cooling to space'; see Section 3.6.3), the cooling rate C is given (in units of K s^{-1}) by

$$C = -\frac{g}{c_p}\left(\frac{d\mathcal{T}^*}{dp}\right)\pi B(T)\,\Delta\nu,$$

where $\mathcal{T}^*(p)$ is the transmittance from pressure level p to space (cf. equation (3.34)). The spectrally integrated Planck function $B(T)\,\Delta\nu$ for the CO_2 band can be approximated by

$$B(T)\,\Delta\nu = 15(T/300)^{3.8}\ \text{W m}^{-2}\,\text{steradian}^{-1}$$

and $\mathcal{T}^*$ can be taken to have the form

$$\mathcal{T}^* = \exp(-\beta p)$$

where $\beta = 3.6 \times 10^{-4}$ Pa^{-1}. Diurnal variations can be crudely accounted for by dividing $h(p_{\text{m}})$ by 2 (why?). Estimate the radiative equilibrium temperature at p_{m}.

3.12. Consider a layer cloud that both absorbs and scatters solar radiation, so that the single-scattering albedo $\omega < 1$. Put $1 - \omega f = a$ and $\omega(1 - f) = b$ and rewrite equations (3.45) as

$$\left(\frac{d}{d\chi^*} + a\right)F^{\downarrow} = bF^{\uparrow}, \qquad \left(\frac{d}{d\chi^*} - a\right)F^{\uparrow} = -bF^{\downarrow}.$$

Hence show that

$$\left(\frac{d^2}{d\chi^{*2}} - \alpha^2\right)F^{\uparrow} = 0, \qquad \text{where} \qquad \alpha^2 = a^2 - b^2.$$

Verify that $\alpha^2 > 0$, so the solution for $F^{\uparrow}$ is a combination of exponentials or, equivalently, of hyperbolic functions. Using one of the boundary conditions (3.48), show that $F^{\uparrow}$ must take the form

$$F^{\uparrow}(\chi^*) \propto \sinh\alpha\,(\chi_1^* - \chi^*).$$

Show that $F^{\downarrow}$ involves both sinh and cosh functions and, by applying the remaining boundary conditions, derive the reflection coefficient R and the transmission coefficient T for the cloud:

$$\mathsf{R} = \frac{b\sinh\theta}{\alpha\cosh\theta + a\sinh\theta}, \qquad \mathsf{T} = \frac{\alpha}{\alpha\cosh\theta + a\sinh\theta},$$

where $\theta = \alpha\chi_1^*$.

Show that, as the optical depth χ_1^* of the layer becomes large,

$$\mathsf{R} \to \mathsf{R}_\infty = \frac{b}{\alpha + a} = \frac{b}{(a^2 - b^2)^{1/2} + a}, \qquad \mathsf{T} \to 0.$$

Evaluate the reflection and transmission coefficients for $\omega = 0.9997$, $f = 0.9$ and $\chi_1^* = 10$, 20 and 30. Note that $\mathsf{R} + \mathsf{T} < 1$: what is the physical reason for this? Find also the reflection coefficient for an infinitely deep cloud with these values of ω and f.

CHAPTER FOUR

Basic fluid dynamics

The atmosphere is a fluid in which a wide variety of flows occurs. This chapter introduces the basic fluid-dynamical laws that govern these atmospheric flows. The length scales of interest range from metres to thousands of kilometres; these are many orders of magnitude greater than molecular scales such as the mean free path, at least in the lower and middle atmosphere. We may therefore average over many molecules, ignoring individual molecular motions and regarding the fluid as continuous. 'Local' values of quantities such as density, temperature and velocity may be defined at length scales that are much greater than the mean free path but much less than the scales on which the meteorological motion varies.

In Section 4.1 we derive the mass-conservation law (often called the *continuity equation*) for a fluid. In Section 4.2 we introduce the concept of the material derivative and the Eulerian and Lagrangian views of fluid motion. An alternative form of the mass conservation law is given in Section 4.3 and the equation of state for the atmosphere (an ideal gas) is recalled in Section 4.4. Then in Section 4.5 Newton's Second Law is applied to a continuous fluid, giving the *Navier–Stokes equation.* The Earth's rotation cannot be ignored for large-scale atmospheric flows, so its incorporation into the Navier–Stokes equation is discussed in Section 4.6. The full equations of motion for a spherical Earth and for Cartesian tangent–plane geometry are given in Section 4.7. Simplifications of these equations for large-scale flows are introduced in Section 4.8. A useful alternative formulation with pressure, rather than height, as a vertical coordinate is briefly mentioned in Section 4.9. Finally, in Section 4.10, we introduce the First Law of Thermodynamics, as applied to the atmosphere. Some applications of these results to atmospheric phenomena are discussed in the problems at the end of the chapter and further applications are provided in Chapter 5.

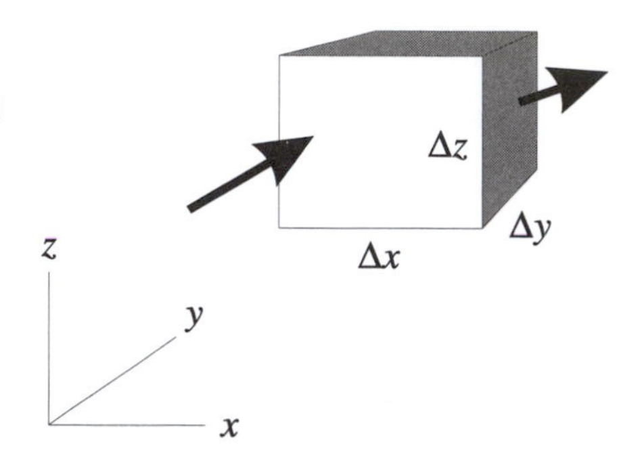

Figure 4.1 A volume element, fixed in space, with fluid passing into and out of it.

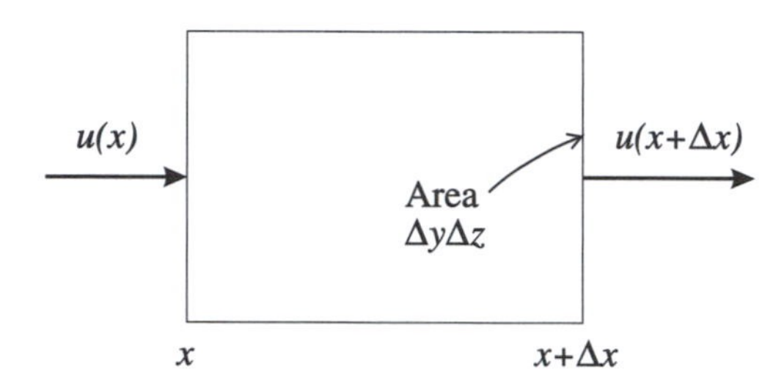

Figure 4.2 A side view of the box in Figure 4.1.

4.1 Mass conservation

Consider a small 'box', or volume element, of sides Δx, Δy and Δz, fixed in space, with fluid passing through it (Figure 4.1). Let the fluid velocity be $\boldsymbol{u} = (u, v, w)$ and the fluid density be ρ; these generally vary with position $\boldsymbol{r}$ and time t.

The net mass inflow in the x-direction per unit time is the difference between that coming in on the left and that going out on the right (see Figure 4.2); i.e.,

$$\Big[\rho(x)u(x) - \rho(x + \Delta x)u(x + \Delta x)\Big]\Delta y\,\Delta z$$
$$\approx -\frac{\partial}{\partial x}(\rho u)\,\Delta x \cdot \Delta y\,\Delta z = -\frac{\partial}{\partial x}(\rho u)\,\Delta V,$$

where $\Delta V = \Delta x\,\Delta y\,\Delta z$, the volume of the box. Including the inflow and outflow in the y- and z-directions, the net inflow is therefore

$$-\nabla \cdot (\rho \boldsymbol{u})\,\Delta V. \tag{4.1}$$

This must equal the rate of increase of mass (density times volume) in the box, namely

$$\frac{\partial}{\partial t}(\rho\,\Delta V). \tag{4.2}$$

However, the box is fixed, so its volume ΔV is constant. Hence, on cancelling ΔV from equations (4.1) and (4.2), we obtain the

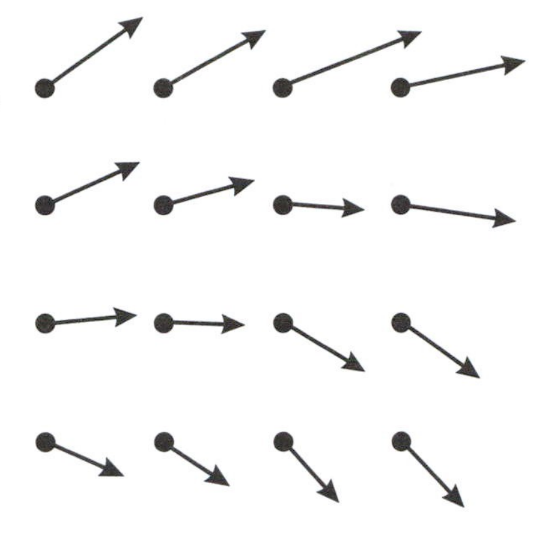

Figure 4.3 A schematic diagram of fluid flow vectors at each point on a grid.

continuity equation, or *mass conservation law*

$$\frac{\partial \rho}{\partial t} + \nabla \cdot (\rho \boldsymbol{u}) = 0. \tag{4.3}$$

By expansion of the $\nabla \cdot (\rho \boldsymbol{u})$ term using a standard vector-calculus identity, this can also be written as

$$\frac{\partial \rho}{\partial t} + \boldsymbol{u} \cdot \nabla \rho + \rho \nabla \cdot \boldsymbol{u} = 0. \tag{4.4}$$

4.2 The material derivative

We can measure flow with respect to fixed points $\boldsymbol{r}$ in space. For example, an anemometer measures wind speed and a wind-vane measures wind direction; together these give us the vector wind $\boldsymbol{u}$. Suppose that such measurements are made for a fixed grid of points, as shown schematically in Figure 4.3; this is the *Eulerian* picture of the flow.

An alternative is to imagine following the motion of tiny 'blobs' of fluid (e.g., marked by a dye). Each blob is supposed to contain many molecules, but moves as a coherent entity, at least for short times; however, in general a blob will be distorted by the flow, as shown in Figure 4.4.

Let the position of a given blob (or rather, its centre of mass) at time t be $\boldsymbol{r}(t)$. The trajectory of its motion between times 0 and t might be as in Figure 4.5. Now consider the kinematics of the blob, ignoring any distortion. Its velocity at time t is the Eulerian velocity $\boldsymbol{u}$ evaluated at the current position $\boldsymbol{r}(t)$ of the blob; this is also equal to the current time rate-of-change of the blob's position. Hence

$$\boldsymbol{u}\big(\boldsymbol{r}(t), t\big) = \frac{d\boldsymbol{r}}{dt}. \tag{4.5}$$

In a similar fashion the acceleration of the blob at time t, say

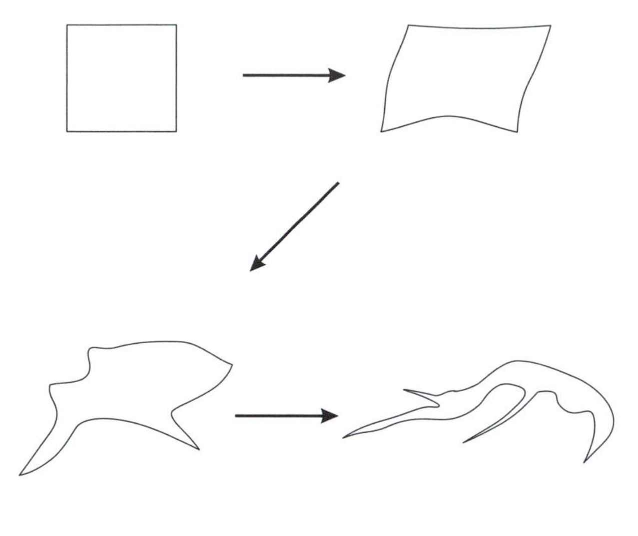

Figure 4.4 A schematic illustration of the possible distortion with time of a moving blob of fluid, initially rectangular in shape.

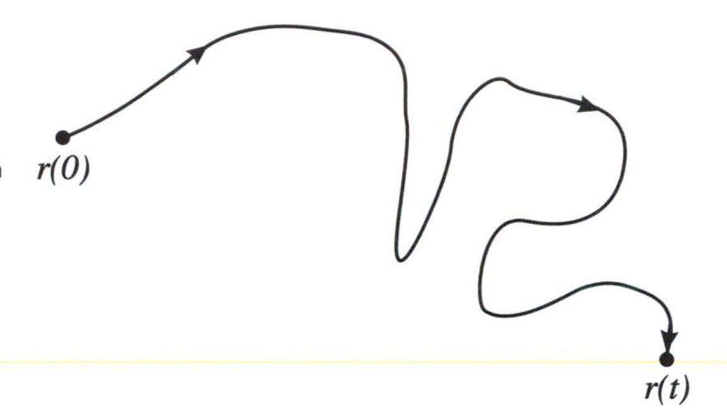

Figure 4.5 A schematic picture of the trajectory of a moving blob of fluid. On this scale, the blob is too small for its distortion in shape to be shown.

$\boldsymbol{a}\big(\boldsymbol{r}(t), t\big)$, equals the second derivative of the position vector:

$$\boldsymbol{a}\big(\boldsymbol{r}(t), t\big) = \frac{d^2\boldsymbol{r}}{dt^2}.$$

This picture of fluid motion in terms of moving blobs is called the *Lagrangian* picture. In practice we may not be able to follow blobs for long; nevertheless, the Lagrangian picture is important conceptually.

The connection between the Lagrangian and Eulerian pictures is made as follows. Let $\boldsymbol{r} = (x, y, z)$ and $\boldsymbol{u} = (u, v, w)$. The x-component of equation (4.5) states that

$$\frac{dx}{dt} = u\big(x(t), y(t), z(t), t\big),$$

so, differentiating with respect to t once again and using the chain rule on the right,

$$\begin{aligned}\frac{d^2x}{dt^2} &= \frac{\partial u}{\partial x}\frac{dx}{dt} + \frac{\partial u}{\partial y}\frac{dy}{dt} + \frac{\partial u}{\partial z}\frac{dz}{dt} + \frac{\partial u}{\partial t} \\ &= \frac{\partial u}{\partial t} + u\frac{\partial u}{\partial x} + v\frac{\partial u}{\partial y} + w\frac{\partial u}{\partial z} \\ &= \frac{\partial u}{\partial t} + \boldsymbol{u}\cdot\nabla u \equiv \frac{Du}{Dt},\end{aligned}$$

(where the three components of equation (4.5) have been used) or in vector form

$$\boldsymbol{a} = \frac{d^2\boldsymbol{r}}{dt^2} = \frac{\partial \boldsymbol{u}}{\partial t} + (\boldsymbol{u}\cdot\nabla)\boldsymbol{u} = \frac{D\boldsymbol{u}}{Dt}. \tag{4.6}$$

We have introduced here the important operator

$$\frac{D}{Dt} \equiv \frac{\partial}{\partial t} + \boldsymbol{u}\cdot\nabla, \tag{4.7}$$

known as the *material derivative* or *advective derivative*. This represents the rate of change with respect to time *following the motion*, (or following a blob) and should be contrasted to $\partial/\partial t$, the rate of change with respect to time *at a fixed point*. (The notation d/dt is also sometimes used for the material derivative in meteorological books.)

Note that the contribution $(\boldsymbol{u}\cdot\nabla)\boldsymbol{u}$ to the material derivative is *nonlinear* in $\boldsymbol{u}$. This makes the behaviour of the atmosphere difficult to forecast (for example, nonlinear systems are well known to display chaotic behaviour), but also leads to many interesting features.

4.3 An alternative form of the continuity equation

Using the definition (4.7), the continuity equation (4.4) can be rewritten as

$$\frac{D\rho}{Dt} + \rho\nabla\cdot\boldsymbol{u} = 0. \tag{4.8}$$

This form of the continuity equation has a useful interpretation in Lagrangian terms, as follows.

Rather than considering a fixed 'box' in space, as above, we focus on a small, moving, initially rectangular 'blob'†. The blob is of fixed mass δm and time-varying volume $\delta V = \delta x\,\delta y\,\delta z$ and travels in a unidirectional flow $(u(x), 0, 0)$ in the x-direction; see Figure 4.6.

† In this chapter we use δ to represent small Lagrangian quantities and Δ to represent small Eulerian quantities. We also use the word 'box' to indicate an infinitesimal volume element *fixed in space* and 'blob' to indicate a *moving* infinitesimal mass element. The latter is similar to the 'parcel' concept introduced in Section 2.5.

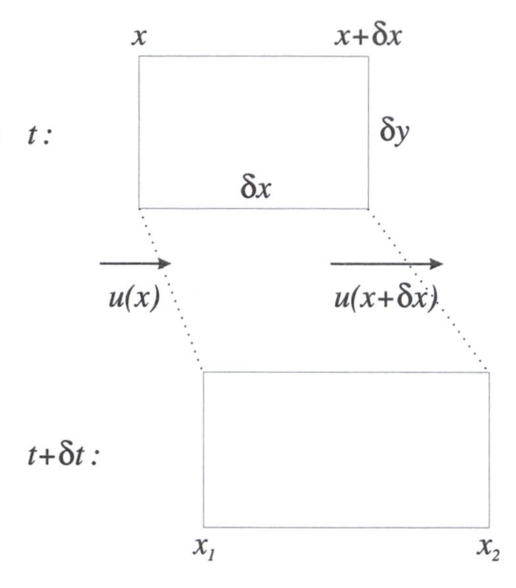

Figure 4.6 A rectangular blob, at times t and $t + \delta t$ in a flow that is moving in the x-direction and also varies in the x-direction.

This flow is assumed to vary in the x-direction; hence, within a small time interval δt the left-hand end of the blob moves from x to $x_1 = x + u(x)\,\delta t$ while the right-hand end moves from $x + \delta x$ to $x_2 = x+\delta x+u(x+\delta x)\,\delta t$. The blob's x-length therefore increases by an amount $[u(x+\delta x)-u(x)]\,\delta t$. Since the velocity vector has no y- or z-component, the dimensions δy and δz remain unchanged. The new volume of the blob is therefore $\delta V' = \{\delta x+[u(x+\delta x)-u(x)]\delta t\}\,\delta y\,\delta z$ and this may be approximated, using a Taylor expansion, by

$$\delta V' \approx \left(\delta x + \frac{\partial u}{\partial x}\,\delta x\,\delta t\right)\delta y\,\delta z = \delta V\left(1 + \frac{\partial u}{\partial x}\,\delta t\right).$$

So

$$\frac{D(\delta V)}{Dt} = \lim_{\delta t\to 0}\left(\frac{\delta V' - \delta V}{\delta t}\right) = (\delta V)\,\frac{\partial u}{\partial x}.$$

More generally, allowing for the flow to be three-dimensional and varying in all directions, so that $\boldsymbol{u} = \boldsymbol{u}(\boldsymbol{r}, t)$, we find that

$$\frac{D(\delta V)}{Dt} = (\delta V)\nabla\cdot\boldsymbol{u}$$

However, $\delta V = \delta m/\rho$ and the constant mass δm can be cancelled, giving

$$\frac{D\rho}{Dt} = -\rho\nabla\cdot\boldsymbol{u},$$

which is equivalent to equation (4.8).

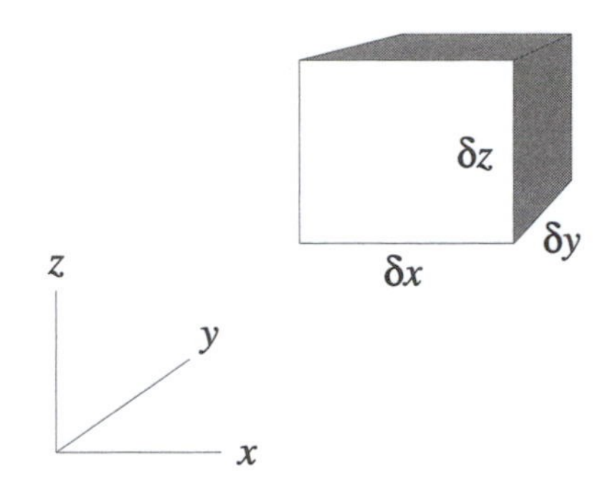

Figure 4.7 An instantaneous view of a small blob of fluid.

4.4 The equation of state for the atmosphere

As in Chapter 2, we assume that the atmosphere is an ideal gas, so that the equation of state is

$$p = RT\rho, \tag{4.9}$$

where p is the pressure, T is the temperature, ρ is the density and R is the gas constant per unit mass of air; cf. equation (2.2).

4.5 The Navier–Stokes equation

In this section we derive the *Navier–Stokes equation* for a fluid, by applying Newton's Second Law to a small moving blob of fluid. Since the blob is moving, this means that we are using the Lagrangian perspective.

Assume that the blob is instantaneously of cuboidal shape, with sides δx, δy and δz, as in Figure 4.7. The volume of the blob is $\delta V = \delta x\, \delta y\, \delta z$ and its mass is $\rho\, \delta V$, where the density is ρ, as above. Newton's Second Law states that

$$(\rho\, \delta V)\boldsymbol{a} = \boldsymbol{F}, \tag{4.10}$$

where $\boldsymbol{a}$ is the acceleration of the blob and $\boldsymbol{F}$ is the vector sum of the forces (including pressure forces, gravity and friction) acting on the blob. We must consider each of these forces in turn.

First, consider the pressure forces acting on the blob and single out the x-direction (see Figure 4.8). At position x the pressure force is $p(x)\, \delta A$ in the positive x-direction, where $\delta A = \delta y\, \delta z$ is the area of the relevant wall of the blob; at position $x + \delta x$ the pressure force is

$$p(x + \delta x)\, \delta A \approx \left(p(x) + \frac{\partial p}{\partial x}\, \delta x + \ldots \right) \delta A,$$

in the negative x-direction, using a Taylor expansion again. The net

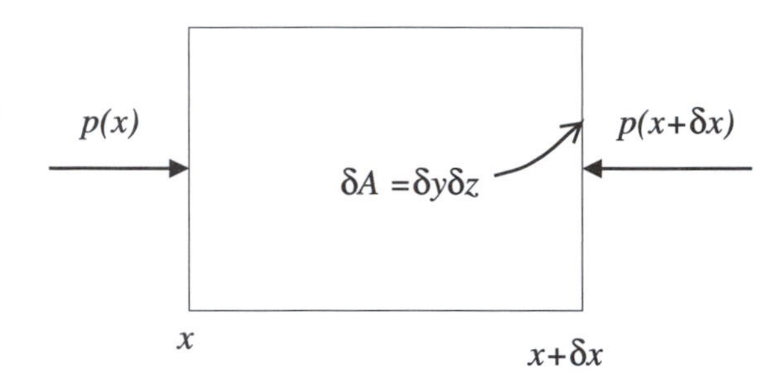

Figure 4.8 Illustrating the various pressure forces acting on the blob in the x-direction.

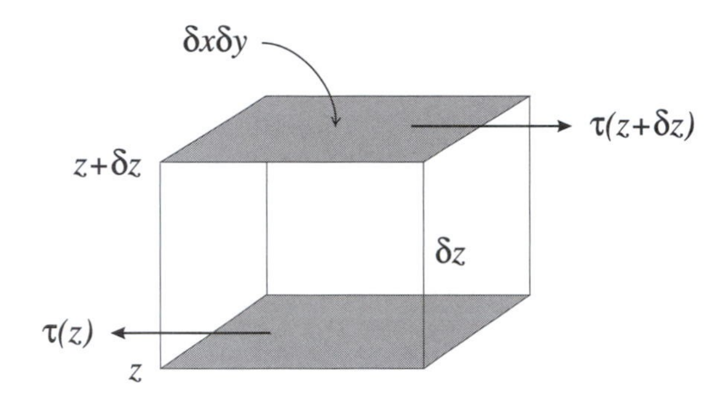

Figure 4.9 Illustrating the horizontal stresses acting on the top and bottom of the blob.

pressure force in the positive x-direction is therefore

$$-\delta x \, \frac{\partial p}{\partial x} \, \delta A = -\delta V \, \frac{\partial p}{\partial x},$$

so in three dimensions the vector pressure force is

$$\boldsymbol{F}_{\text{press}} = -(\delta V)(\nabla p). \tag{4.11}$$

Next consider the gravity force acting on the blob: this is just the mass $\rho \, \delta V$ of the blob times g, acting downwards:

$$\boldsymbol{F}_{\text{grav}} = -(\rho \, \delta V) g \boldsymbol{k}, \tag{4.12}$$

where $\boldsymbol{k}$ is the unit vertical vector, $(0, 0, 1)$.

Finally, consider the viscous forces acting on the blob. These require quite a complicated treatment in general, so we shall consider a special case by way of illustration. Suppose first that the bulk flow (averaged over many molecules) is in the x-direction only, and varies only in the z-direction, i.e., $\boldsymbol{u} = (u(z), 0, 0)$. Recall, from the kinetic theory of gases, that the viscous stress (the force per unit *horizontal* area, in this case) in the x-direction is

$$\tau = \eta \, \frac{du}{dz}, \tag{4.13}$$

where η is the *dynamic viscosity*. (It is usually called μ in fluid dynamics texts.) Now slightly different stresses act on the top and bottom surfaces of our blob, as illustrated in Figure 4.9. The net

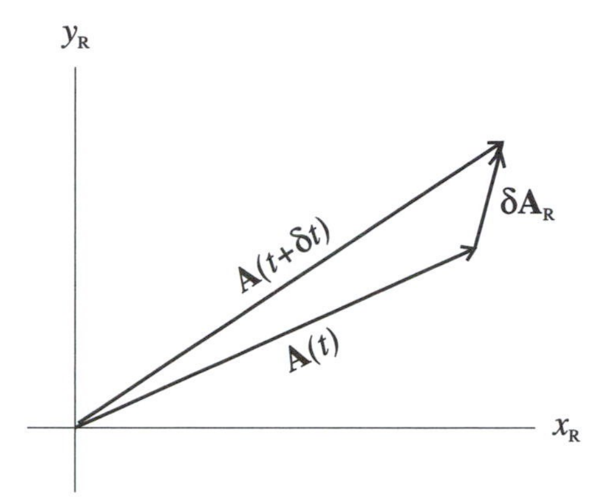

Figure 4.10 The change of the vector **A** viewed in the rotating frame R.

contribution to the viscous force on the blob is the difference between the stresses on the top and bottom, namely

$$\left[\tau(z+\delta z)-\tau(z)\right]\delta x\,\delta y \approx \frac{\partial \tau}{\partial z}\,\delta x\,\delta y\,\delta z = \frac{\partial \tau}{\partial z}\,\delta V; \tag{4.14}$$

using equation (4.13) and assuming that η is constant, this becomes

$$\eta\,\frac{d^2u}{dz^2}\,\delta V. \tag{4.15}$$

The situation is more complicated if there are y- and z-variations as well, but it turns out that the viscous vector force is

$$\boldsymbol{F}_{\text{visc}} = \delta V\,\eta\nabla^2\boldsymbol{u}. \tag{4.16}$$

This is clearly a generalisation of equation (4.15).

Assuming that these are the only relevant forces acting on the blob, putting expressions (4.11), (4.12) and (4.16) into Newton's Second Law (4.10) and using equation (4.6) for the acceleration $\boldsymbol{a}$, we get

$$\frac{D\boldsymbol{u}}{Dt} = -\frac{1}{\rho}\nabla p - g\boldsymbol{k} + \frac{\eta}{\rho}\nabla^2\boldsymbol{u}. \tag{4.17}$$

This is the *Navier–Stokes equation* (or *momentum equation*) for fluid flow in an inertial frame.

4.6 Rotating frames of reference

The rotation of the Earth is significant for the large-scale dynamics of the atmosphere. It is usually most convenient to work with a coordinate system fixed with respect to the Earth; since this system is rotating with respect to inertial space, modifications to the Navier–Stokes equation (4.17) must be made.

These modifications involve changes to the time derivatives of vectors. Suppose that the frame R rotates at a constant angular

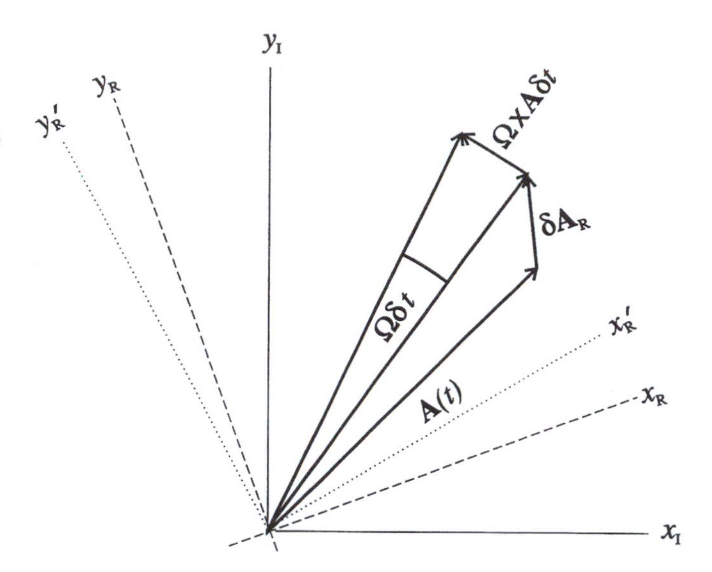

Figure 4.11 The change of the vector A viewed in the inertial frame I. In time δt the axes fixed in R, as observed in I, rotate through an angle $\Omega\,\delta t$, from $x_R y_R$ to $x'_R y'_R$.

velocity $\boldsymbol{\Omega}$ with respect to an inertial frame I and define the z-axes of both frames to be in the direction of $\boldsymbol{\Omega}$. Now consider a time-varying vector $\boldsymbol{A}(t)$ and suppose, for simplicity, that it lies in the common x, y plane of R and I. As viewed in the rotating frame R, $\boldsymbol{A}$ changes from $\boldsymbol{A}(t)$ to $\boldsymbol{A}(t+\delta t) \approx \boldsymbol{A}(t) + \delta \boldsymbol{A}_{\mathrm{R}}$, in the time interval between t and $t+\delta t$, where

$$\delta \boldsymbol{A}_{\mathrm{R}} \equiv \left(\frac{d\boldsymbol{A}}{dt}\right)_{\mathrm{R}} \delta t,$$

as illustrated in Figure 4.10. However, if the same vector $\boldsymbol{A}$ is viewed in the inertial frame I, we must allow for the rotation of the frame R with respect to the frame I in the time δt; this gives an extra contribution to the change in $\boldsymbol{A}$ when it is viewed in the inertial frame, as shown in Figure 4.11. With a little consideration of vector geometry, the extra contribution to the change in $\boldsymbol{A}$ is found to be $\boldsymbol{\Omega} \times \boldsymbol{A}\,\delta t$, for small δt. (The same can be shown to hold if $\boldsymbol{A}$ is not in the x, y plane.) So, as $\delta t \to 0$, we obtain the following relationship between time derivatives in the inertial frame I and the rotating frame R:

$$\left(\frac{d\boldsymbol{A}}{dt}\right)_{\mathrm{I}} = \left(\frac{d\boldsymbol{A}}{dt}\right)_{\mathrm{R}} + \boldsymbol{\Omega} \times \boldsymbol{A}. \tag{4.18}$$

A double application of equation (4.18) gives

$$\left(\frac{d^2\boldsymbol{A}}{dt^2}\right)_{\mathrm{I}} = \left(\frac{d^2\boldsymbol{A}}{dt^2}\right)_{\mathrm{R}} + 2\boldsymbol{\Omega} \times \left(\frac{d\boldsymbol{A}}{dt}\right)_{\mathrm{R}} + \boldsymbol{\Omega} \times (\boldsymbol{\Omega} \times \boldsymbol{A}).$$

In particular, if $\boldsymbol{A} = \boldsymbol{r}$, the position vector, we obtain the following relationship between the acceleration in the inertial frame and the

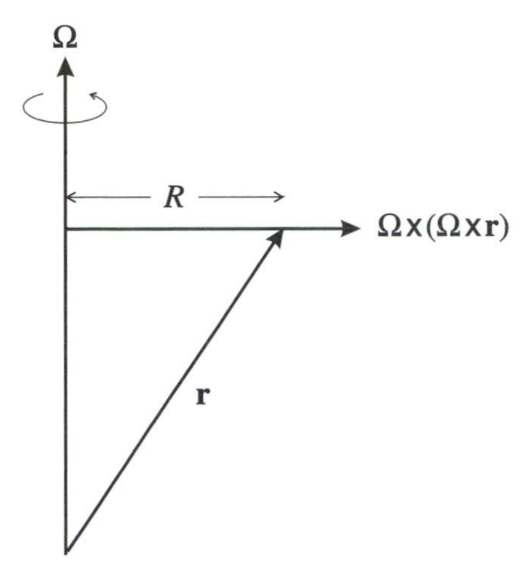

Figure 4.12 Illustrating the centripetal acceleration.

acceleration, velocity and position vector in the rotating frame:

$$\boldsymbol{a}_{\mathrm{I}} = \boldsymbol{a}_{\mathrm{R}} + 2\boldsymbol{\Omega} \times \boldsymbol{u}_{\mathrm{R}} + \boldsymbol{\Omega} \times (\boldsymbol{\Omega} \times \boldsymbol{r}). \tag{4.19}$$

Let us examine the terms involving $\boldsymbol{\Omega}$ on the right-hand side of equation (4.19). The term $2\boldsymbol{\Omega} \times \boldsymbol{u}_{\mathrm{R}}$ represents the *Coriolis acceleration*: it is perpendicular both to the velocity $\boldsymbol{u}_{\mathrm{R}}$ in the rotating frame and to $\boldsymbol{\Omega}$. The term $\boldsymbol{\Omega} \times (\boldsymbol{\Omega} \times \boldsymbol{r})$ represents the *centripetal acceleration*: it has magnitude $\Omega^2 R$, where R is the perpendicular distance from the point $\boldsymbol{r}$ to the rotation axis, and is directed perpendicularly away from the rotation axis, as shown in Figure 4.12.

To rewrite the Navier–Stokes equation in the rotating frame, we need to replace the acceleration $\boldsymbol{a} = D\boldsymbol{u}/Dt$ by $\boldsymbol{a}_{\mathrm{I}}$ in equation (4.17), to get the corresponding equation *with respect to the rotating frame*:

$$\frac{D\boldsymbol{u}}{Dt} = -\frac{1}{\rho}\nabla p - 2\boldsymbol{\Omega} \times \boldsymbol{u} - \boldsymbol{\Omega} \times (\boldsymbol{\Omega} \times \boldsymbol{r}) - g\boldsymbol{k} + \frac{\eta}{\rho}\nabla^2 \boldsymbol{u}, \tag{4.20}$$

where the subscript R has been dropped and $\boldsymbol{u}$ now represents the velocity measured in the rotating frame R. With the acceleration terms $2\boldsymbol{\Omega} \times \boldsymbol{u}$ and $\boldsymbol{\Omega} \times (\boldsymbol{\Omega} \times \boldsymbol{r})$ written on the right-hand side of equation (4.20), they can be regarded as *fictitious forces* in Newton's Second Law: the *Coriolis force* and *centrifugal force*[†], respectively. We often put $\boldsymbol{g}' = -g\boldsymbol{k} - \boldsymbol{\Omega} \times (\boldsymbol{\Omega} \times \boldsymbol{r})$, the *effective gravity*. The extra term is much less than g in magnitude and is in a different direction.

[†] Note that the acceleration is centripetal ('centre-seeking') whereas the fictitious force, being of opposite sign, is centrifugal ('centre-fleeing').

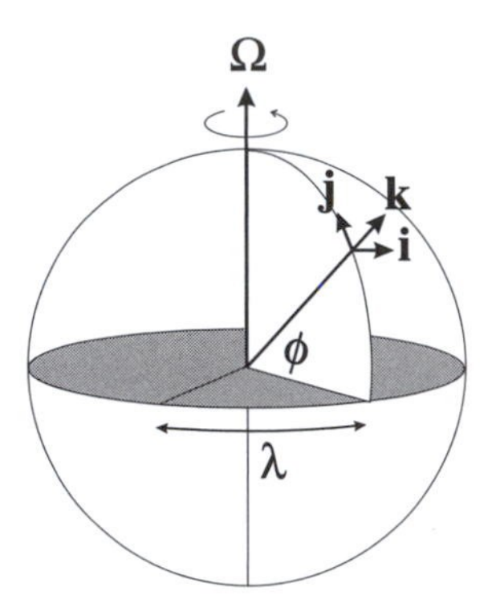

Figure 4.13 Local Cartesian coordinate axes on a spherical Earth.

4.7 Equations of motion in coordinate form

4.7.1 Spherical coordinates

The natural coordinates in which to express our equations, when they are applied to the Earth, are spherical coordinates† (r, ϕ, λ), where r is the distance from the centre of the Earth, ϕ is latitude and λ is longitude. At a point on the Earth's surface we draw unit vectors $\boldsymbol{i}$ pointing eastwards, $\boldsymbol{j}$ pointing northwards and $\boldsymbol{k}$ pointing upwards, as shown in Figure 4.13.

It is convenient to introduce small incremental distances $dx = r\cos\phi\, d\lambda$ in the eastward (or *zonal*) direction and $dy = r\,d\phi$ in the northward (or *meridional*) direction. We also introduce the vertical distance z from the Earth's surface, so that $r = a + z$, where a is the Earth's radius and $dz = dr$.

It is shown in Appendix B that, neglecting the centripetal acceleration, equation (4.20) can be written in component form as follows:

$$\frac{Du}{Dt} - \left(2\Omega + \frac{u}{r\cos\phi}\right)(v\sin\phi - w\cos\phi) + \frac{1}{\rho}\frac{\partial p}{\partial x} = F^{(x)}, \quad (4.21a)$$

$$\frac{Dv}{Dt} + \frac{wv}{r} + \left(2\Omega + \frac{u}{r\cos\phi}\right)u\sin\phi + \frac{1}{\rho}\frac{\partial p}{\partial y} = F^{(y)}, \quad (4.21b)$$

$$\frac{Dw}{Dt} - \frac{u^2 + v^2}{r} - 2\Omega u\cos\phi + \frac{1}{\rho}\frac{\partial p}{\partial z} + g = F^{(z)}, \quad (4.21c)$$

where $F^{(x)}$, $F^{(y)}$ and $F^{(z)}$ are components of the frictional force in the eastward, northward and upward directions, respectively, and

$$\frac{D}{Dt} = \frac{\partial}{\partial t} + u\frac{\partial}{\partial x} + v\frac{\partial}{\partial y} + w\frac{\partial}{\partial z}. \quad (4.22)$$

† Strictly speaking, oblate spheroidal coordinates should be used; see Gill (1982). We shall ignore this complication here.

4.7.2 Approximations to the spherical equations

Equations (4.21) are complicated, but approximate versions are sufficient for modelling many atmospheric dynamical phenomena. In the first place, we can replace the distance r by the Earth's radius a with negligible error, since the part of the atmosphere in which we are interested has a depth of 100 km or so, which is much less than $a \approx 6400$ km.

Having done this, consider equation (4.21a). Two useful simplifications can be made here.

- The zonal wind will generally be less than $100\,\mathrm{m\,s^{-1}}$ in magnitude. It can then be verified that

$$\frac{|u|}{a \cos \phi} \ll 2\Omega,$$

  except perhaps near the poles, where $\cos\phi \to 0$. (In this calculation we can use the fact that Ωa, the tangential speed of the Earth's surface at the equator, is approximately $465\,\mathrm{m\,s^{-1}}$.)
- Vertical velocities are usually much less than horizontal velocities, so

$$|w \cos \phi| \ll |v \sin \phi|,$$

  except perhaps near the equator, where $\sin\phi \to 0$.

Given these two results, and introducing the *Coriolis parameter*,

$$f = 2\Omega \sin \phi,$$

equation (4.21a) reduces to

$$\frac{Du}{Dt} - fv + \frac{1}{\rho}\frac{\partial p}{\partial x} = F^{(x)}. \tag{4.23a}$$

Equation (4.21b) can be simplified in a similar way (but using also the result that $|wv|/r \ll 2\Omega|u \sin\phi|$, except possibly near the equator), to give

$$\frac{Dv}{Dt} + fu + \frac{1}{\rho}\frac{\partial p}{\partial y} = F^{(y)}. \tag{4.23b}$$

In equation (4.21c) it is easy to show that the terms $(u^2 + v^2)/a$ and $2\Omega u \cos\phi$ are very much smaller than $g = 9.8\,\mathrm{m\,s^{-2}}$ for any reasonable values of the horizontal velocity components u and v. Omitting also the vertical friction term $F^{(z)}$, which is usually regarded as negligible, we obtain

$$\frac{Dw}{Dt} + \frac{1}{\rho}\frac{\partial p}{\partial z} + g = 0. \tag{4.23c}$$

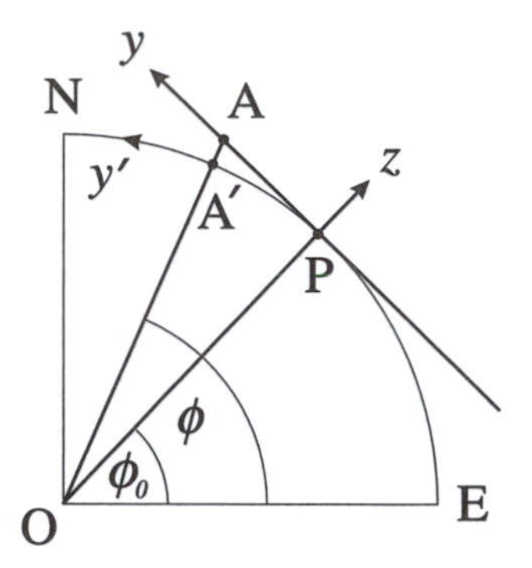

Figure 4.14 Illustrating the use of tangent-plane geometry. The figure shows a section through the centre of the Earth, O, the North Pole, N, and the point P (at latitude ϕ_0); E denotes a point on the Equator. The plane AP is tangential to the Earth at point P. Northward distances y are measured in this plane, whereas northward distances y' are measured on the surface of the Earth. These distances are almost equal if the latitude ϕ of point A′ on the surface of the Earth is close to ϕ_0.

4.7.3 Tangent-plane geometry

Equations (4.23a)–(4.23c) are expressed in spherical coordinates. However, the use of dx and dy as small eastward and northward distances suggests a further useful simplification, which is valid when we are considering a comparatively small region near a point P at latitude ϕ_0 and longitude λ_0. In this case we can introduce Cartesian coordinates (x, y, z) on the *tangent plane* at the point P: clearly there is little difference between distances (x', y', say) on the surface of the Earth and distances (x, y) on the tangent plane in the neighbourhood of P (see Figure 4.14). We can therefore re-interpret equations (4.23a)–(4.23c) as applying to the Cartesian coordinates on the tangent plane and avoid complications due to spherical geometry. We must, however, replace f (which varies with latitude) by the constant value

$$f_0 = 2\Omega \sin \phi_0.$$

Note that equations (4.23a) and (4.23b) then become identical† to those for a system that is rotating about the z-axis with angular velocity $\frac{1}{2} f_0$. This approximation is called *the f-plane approximation* and the analogous system is *the f-plane.*

If we wish to consider a larger region we may retain the tangent-plane approximation but allow for some variation of f with latitude. A Taylor expansion of $f(\phi)$ about $\phi = \phi_0$ gives

$$f(\phi) = 2\Omega \sin \phi = 2\Omega[\sin \phi_0 + (\phi - \phi_0) \cos \phi_0 + \ldots]$$

so that on the tangent plane, where $\phi - \phi_0 \approx y/a$,

$$f(y) \approx f_0 + \beta y, \qquad \text{where} \qquad \beta = \frac{2\Omega \cos \phi_0}{a} = \left(\frac{df}{dy}\right)_{y=0}. \tag{4.24}$$

Equation (4.24) is called *the β-plane approximation*: instead of taking f to be constant, as in the f-plane approximation, we allow it to vary linearly with the northward distance y.

† Except that centripetal accelerations are again neglected.

Scale	*Symbol*	*Typical magnitude*
Horizontal scale	L	$1000\,\mathrm{km} = 10^6$ m
Vertical scale	H	$10\,\mathrm{km} = 10^4$ m
Horizontal velocity	U	$10\,\mathrm{m\,s^{-1}}$
Vertical velocity	W	$10^{-2}\,\mathrm{m\,s^{-1}}$
Timescale	T	1 day $\sim 10^5$ s
Surface density	ρ	$1\,\mathrm{kg\,m^{-3}}$
Earth's radius	a	6.4×10^6 m
2 × rotation rate	2Ω	$10^{-4}\,\mathrm{s^{-1}}$
Acceleration of gravity	g	$10\,\mathrm{m\,s^{-2}}$

Table 4.1. *Some scales for large-scale motion in the atmosphere.*

4.8 Geostrophic and hydrostatic approximations

Under appropriate dynamical conditions we can simplify equations (4.23a)–(4.23c) still further, using the method of *scale analysis.* (A simple form of scale analysis was used in Section 4.7.2 to simplify equations (4.21a)–(4.21c).) For example, consider motions associated with *synoptic-scale* systems – that is, large-scale weather systems – at midlatitudes, with the time and space scales given in Table 4.1. First consider typical sizes of the terms in the vertical momentum equation (4.23c). The material derivative can be expanded as

$$\frac{Dw}{Dt} = \frac{\partial w}{\partial t} + u\frac{\partial w}{\partial x} + v\frac{\partial w}{\partial y} + w\frac{\partial w}{\partial z}.$$

Very roughly, we can estimate the individual terms here as

$$\frac{\partial w}{\partial t} \sim \frac{W}{T} \sim \frac{10^{-2}}{10^5} \sim 10^{-7}\,\mathrm{m\,s^{-2}},$$

$$u\frac{\partial w}{\partial x} + v\frac{\partial w}{\partial y} \sim \frac{UW}{L} \sim \frac{10^{-1}}{10^6} \sim 10^{-7}\,\mathrm{m\,s^{-2}},$$

$$w\frac{\partial w}{\partial z} \sim \frac{W^2}{H} \sim \frac{10^{-4}}{10^4} \sim 10^{-8}\,\mathrm{m\,s^{-2}};$$

hence in total we estimate

$$\frac{Dw}{Dt} \sim 10^{-7}\,\mathrm{m\,s^{-2}}.$$

This is very much smaller than the g term in equation (4.23c):

$$g \sim 10^1\,\mathrm{m\,s^{-2}}.$$

The remaining term is $(1/\rho)\,\partial p/\partial z$; this is the only term that can possibly balance the large g term, so equation (4.21c) must become, to a good approximation,

$$\frac{\partial p}{\partial z} = -g\rho. \tag{4.25}$$

This shows that, under our assumed scaling, the vertical momentum equation reduces to *hydrostatic balance* (cf. Chapter 2, equation (2.12)).

Now let us perform a similar scale analysis on the horizontal momentum equation (4.23a):

$$\frac{\partial u}{\partial t} \sim 10^{-4}, \qquad u\frac{\partial u}{\partial x} \sim 10^{-4}, \qquad w\frac{\partial u}{\partial z} \sim 10^{-5},$$

$$fv = 2\Omega v \sin\phi \sim 10^{-3}$$

(all in units of $\mathrm{m\,s^{-2}}$), where a midlatitude value of ϕ is used and $F^{(x)}$ is assumed negligible.

Here fv is the biggest term (just), so it must be balanced by the remaining term $(1/\rho)\,\partial p/\partial x$, giving the *geostrophic approximation*:

$$fv = \frac{1}{\rho}\frac{\partial p}{\partial x}; \tag{4.26a}$$

a similar scale analysis applied to equation (4.21b) yields

$$-fu = \frac{1}{\rho}\frac{\partial p}{\partial y}. \tag{4.26b}$$

Thus, for synoptic-scale systems, the horizontal momentum equations (4.23a) and (4.23b) reduce to *geostrophic balance*, in which the horizontal pressure gradients are balanced by Coriolis forces associated with the horizontal winds. The following points should be noted.

- At each height the horizontal wind blows along the *isobars* (the lines of constant pressure), since $(u, v, 0)\cdot\nabla p = 0$ under the geostrophic approximation.
- It is twice the *vertical* component of the rotation vector, $2\Omega\sin\phi$, that enters.
- The ratio of the 'horizontal advection' term to the 'Coriolis term' in equation (4.21a) is approximately

$$\frac{u\,\partial u/\partial x}{fv} \sim \frac{U^2/L}{fU} = \frac{U}{fL} \equiv \mathrm{Ro}; \tag{4.27}$$

  Ro is a dimensionless number, called the *Rossby Number*. If $\mathrm{Ro} \gg 1$, the Coriolis term is comparatively small, whereas if $\mathrm{Ro} \ll 1$, the Coriolis term is comparatively large and the geostrophic approximation is usually valid.

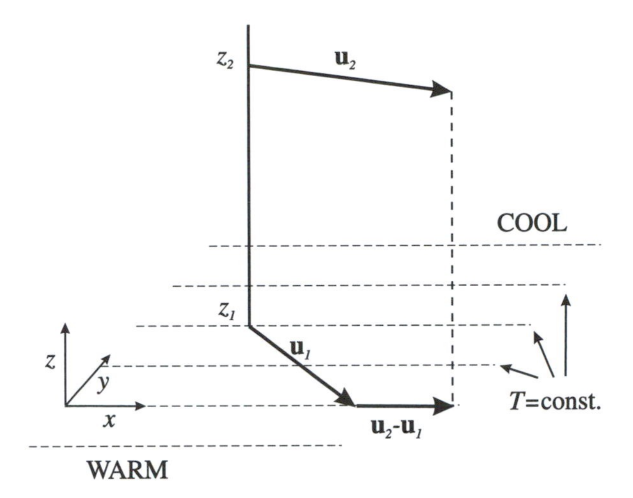

Figure 4.15 Illustrating thermal windshear balance.

4.8.1 The thermal windshear equations

From the geostrophic equation (4.26a) and the ideal gas law (4.9) we have

$$fv = \frac{RT}{p}\frac{\partial p}{\partial x} = RT\frac{\partial \ln p}{\partial x},$$

while from equations (4.25) and (4.9) we have

$$-\frac{g}{RT} = \frac{\partial \ln p}{\partial z}.$$

Neglecting vertical (but not horizontal) variations† in T, cross-differentiation of these gives

$$f\frac{\partial v}{\partial z} \approx \frac{g}{T}\frac{\partial T}{\partial x} \tag{4.28a}$$

and similarly, from equations (4.26b), (4.25) and (4.9), we obtain

$$f\frac{\partial u}{\partial z} \approx -\frac{g}{T}\frac{\partial T}{\partial y}. \tag{4.28b}$$

These are called the *thermal wind equations* or, more correctly, the *thermal windshear equations*; they give a very useful relation between horizontal temperature gradients and vertical gradients of the horizontal wind, when both geostrophic balance and hydrostatic balance apply.

As an example of the use of the thermal windshear equations, consider a case in which the temperature varies only with the northward distance y, so $T = T(y)$, and decreases with y, so $dT/dy < 0$; suppose that the horizontal wind direction varies between $\boldsymbol{u}_1 = (u_1, v_1)$ at a lower level $z = z_1$ and $\boldsymbol{u}_2 = (u_2, v_2)$

† Vertical variations of T need not be neglected if we use pressure p, rather than height z, as a vertical coordinate; see Section 4.9.

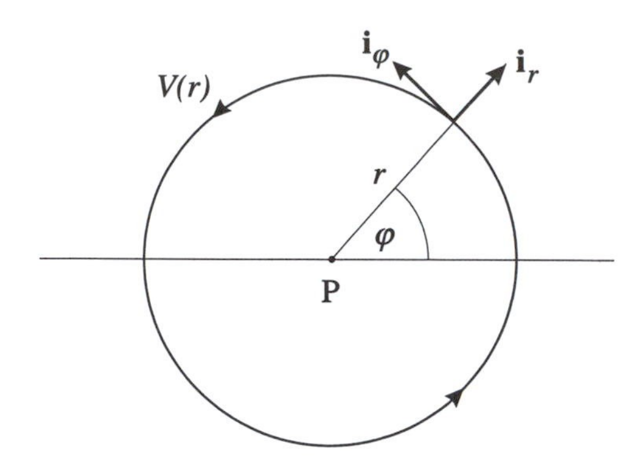

Figure 4.16 Illustrating the circular vortex.

at a higher level $z = z_2$, as shown in Figure 4.15. Since T is independent of x, the isotherms $T =$ constant are parallel to the x-axis; moreover, equation (4.28a) then implies that $\partial v/\partial z = 0$, so $v_2 = v_1$. Therefore the difference between the wind vectors at z_1 and z_2 is parallel to the x-axis and hence to the isotherms $T =$ constant. Given that $dT/dy < 0$, equation (4.28b) implies that $\partial u/\partial z > 0$, so $u_2 > u_1$. It can then be seen that if the wind turns anticlockwise† with height (known as *backing*), the wind on average blows across the isotherms from the cold to the warm side: this is called *cold advection.* Conversely, if the wind turns clockwise with height (*veering*), the wind on average blows across the isotherms from warm to cold (*warm advection*).

Another example is the zonally averaged (i.e., longitudinally averaged) zonal (i.e., east–west) winds in the lower and middle atmosphere, shown in Figure 1.6, which tend to be nearly in thermal-windshear balance with the zonally averaged temperature, shown in Figure 1.5. It can readily be checked that the signs, at least, of $\partial u/\partial z$ and $\partial T/\partial y$ in these figures are consistent with equation (4.28b).

There are many other applications of geostrophic and hydrostatic balance in the interpretation of atmospheric processes. An important case is the structure of fronts; see Problem 4.7.

4.8.2 A circular vortex: gradient–wind balance

We now consider another important type of approximation to the horizontal momentum equations, which is more accurate than geostrophic balance. In fact it is *exact* for purely horizontal flow in the form of a steady vortex with circular streamlines. We consider such a vortex, centred at a point P, near latitude ϕ, and use polar coordinates r and φ, in the horizontal plane, also centred at P, as shown in Figure 4.16. For this circular vortex the velocity vector is $\boldsymbol{u} = V(r)\,\boldsymbol{i}_\varphi$, where $\boldsymbol{i}_\varphi$ is the unit vector in the azimuthal direction.

† In the Northern Hemisphere, where $f > 0$.

The corresponding Coriolis force is $-\rho f \boldsymbol{k} \times \boldsymbol{u} = \rho f V \boldsymbol{i}_r$, where $\boldsymbol{i}_r$ is the unit vector in the radial direction. We also include the centrifugal force $(\rho V^2/r)\boldsymbol{i}_r$. The sum of these two forces has to be balanced by a radial pressure gradient, so

$$\frac{V^2}{r} + fV = \frac{1}{\rho}\frac{dp}{dr} \equiv G,$$

say. This equation expresses *gradient–wind balance* and should be contrasted with geostrophic balance, in which the V^2/r term is neglected. It is a quadratic equation for V, if G is given, with solutions

$$V(r) = -\frac{rf}{2} \pm \left(\frac{r^2 f^2}{4} + Gr\right)^{1/2}.$$

Note that physically sensible solutions are possible only if V is real, i.e., if

$$G = \frac{1}{\rho}\frac{dp}{dr} \geq -rf^2/4,$$

so the pressure cannot drop too rapidly with radius if gradient–wind balance is to occur.

4.9 Pressure coordinates and geopotential

Meteorologists often use pressure p, rather than height z, as a vertical coordinate. This has two, quite separate, advantages: first, atmospheric measurements are usually referenced with respect to pressure rather than height; and second, as we shall see, the equations of motion take a simpler form in these 'pressure coordinates' than they do in height coordinates. However, care needs to be exercised when taking partial derivatives: we need to be clear about *which* variables are held constant in the partial differentiation!

For a small column of air, $dp = -g\rho\, dz$ from hydrostatic balance (equation (4.25)), so the pressure decreases monotonically with height. There is thus a one-to-one relationship between pressure and height, which means that no ambiguity will arise in using p as a vertical coordinate.

We can turn equation (4.25) upside down and use the ideal-gas law, equation (4.9), to get

$$g\left(\frac{\partial z}{\partial p}\right)_{x,y,t} = -\left(\frac{1}{\rho}\right) = -\frac{RT}{p},$$

so, if the *geopotential* $\Phi = gz$ is introduced, the hydrostatic equation becomes

$$\frac{\partial \Phi}{\partial p} = -\frac{RT}{p} \qquad (4.29)$$

in pressure coordinates. In equation (4.29) the subscripts x, y and t (indicating which variables are held constant when the partial p-derivative of Φ is taken) are omitted, since there should be no danger of confusion.

We can derive the pressure-coordinate versions of geostrophic balance as follows. We first eliminate the density ρ from equations (4.25) and (4.26a):

$$fv = -g\left(\frac{\partial p}{\partial x}\right) \Big/ \left(\frac{\partial p}{\partial z}\right) = -g\left(\frac{\partial p}{\partial x}\right)\left(\frac{\partial z}{\partial p}\right),$$

where we have again turned $(\partial p/\partial z)_{x,y,t}$ upside down. We next use the well-known 'reciprocity' theorem of partial differentiation (used, e.g., in thermodynamics):

$$\left(\frac{\partial \alpha}{\partial \beta}\right)_\gamma \left(\frac{\partial \beta}{\partial \gamma}\right)_\alpha \left(\frac{\partial \gamma}{\partial \alpha}\right)_\beta = -1. \tag{4.30}$$

This gives

$$fv = g\left(\frac{\partial z}{\partial x}\right)_{y,p,t} = \left(\frac{\partial \Phi}{\partial x}\right)_{y,p,t} \tag{4.31}$$

and similarly

$$fu = -\left(\frac{\partial \Phi}{\partial y}\right)_{x,p,t}. \tag{4.32}$$

Equations (4.31) and (4.32) are the equations for geostrophic balance in pressure coordinates. A p-coordinate version of the thermal windshear equations can be derived, without assuming that T is independent of z; see Problem 4.8.

The full equations of motion (4.21a)–(4.21c) can be derived in pressure coordinates, but these will not be needed here. Note, however, that instead of the geometric vertical velocity w we must introduce a vertical 'pressure velocity'

$$\omega \equiv \frac{Dp}{Dt},$$

which can be shown to be approximately equal to $-\rho g w$.

In pressure coordinates the equation of continuity of mass, equation (4.4) or (4.8), is replaced by the simpler form

$$\left(\frac{\partial u}{\partial x}\right)_{y,p,t} + \left(\frac{\partial v}{\partial y}\right)_{x,p,t} + \left(\frac{\partial \omega}{\partial p}\right)_{x,y,t} = 0, \tag{4.33}$$

with no time derivative or density ρ appearing. This can be obtained by considering a small moving blob of fluid, as in Figure 4.7, with volume $\delta V = \delta x\,\delta y\,\delta z$ and mass $\delta m = \rho\,\delta V = -\delta x\,\delta y\,\delta p/g$ (using the hydrostatic equation in the form $\delta p = -g\rho\,\delta z$). This mass is constant, following the motion of the blob, so $D(\delta m)/Dt = 0$ and hence

$$0 = \frac{1}{\delta m}\frac{D}{Dt}(\delta m) = \frac{g}{\delta x\,\delta y\,\delta p}\frac{D}{Dt}\left(\frac{\delta x\,\delta y\,\delta p}{g}\right).$$

Using the chain rule and the fact that $D(\delta x)/Dt = D(x_2 - x_1)/Dt = u_2 - u_1 = \delta u$, $D(\delta p)/Dt = D(p_2 - p_1)/Dt = \omega_2 - \omega_1 = \delta\omega$, etc., we get

$$\frac{\delta u}{\delta x} + \frac{\delta v}{\delta y} + \frac{\delta\omega}{\delta p} = 0,$$

which gives equation (4.33) as δx, δy and δp tend to zero.

4.10 The thermodynamic energy equation

The First Law of Thermodynamics for a moving blob of fluid of unit mass, undergoing reversible changes δS of entropy, δU of internal energy and δV of volume in time δt can be written:

$$T\,\delta S = \delta U + p\,\delta V;$$

cf. Chapter 2, equation (2.17). As shown in equation (2.22), this can be written, for an ideal gas, as

$$T\,\delta S = c_p\,\delta T - \frac{1}{\rho}\,\delta p, \tag{4.34}$$

where c_p is the specific heat capacity at constant pressure.

Dividing equation (4.34) by δt, letting $\delta t \to 0$ and remembering that we are considering a moving (Lagrangian) mass, so that the time derivative is D/Dt, not $\partial/\partial t$, we then obtain

$$Q \equiv T\frac{DS}{Dt} = c_p\frac{DT}{Dt} - \frac{1}{\rho}\frac{Dp}{Dt}. \tag{4.35}$$

This is one form of the First Law of Thermodynamics, as used in atmospheric physics. (It is also commonly called the thermodynamic energy equation in meteorology.) The quantity Q, defined by the first member of equation (4.35), is called the *diabatic heating rate per unit mass*. For a blob in *adiabatic* motion, $Q = 0^{\dagger}$.

The main physical processes contributing to Q in the lower and middle atmosphere are latent heating and cooling (from condensation and evaporation, respectively, of water vapour) and radiative heating and cooling (from absorption and emission of electromagnetic radiation; see Chapter 3). Note that, if $Q = 0$,

$$\frac{DT}{Dt} = \frac{1}{\rho c_p}\frac{Dp}{Dt}, \tag{4.36}$$

† The use of Q here is quite common in atmospheric physics; however, it would be more consistent with the notation of Chapter 2, e.g., equation (2.16), to use $\lim(\delta Q/\delta t) = DQ/Dt$ for the diabatic heating rate.

so, in the absence of diabatic heating, the temperature of a moving blob of air will increase if it *descends* (moves to higher pressure) and will decrease if it *ascends*. Another way of showing this is to note that, if the blob moves *adiabatically*, it conserves its *potential temperature*, θ:

$$\theta = T\left(\frac{p_0}{p}\right)^{\kappa},$$

where $\kappa = R/c_p$; see equation (2.25). As p increases, T has to increase as well, to keep θ constant, which is consistent with equation (4.36).

It is frequently more convenient to work with potential temperature θ than with temperature T; in the presence of diabatic heating, the thermodynamic energy equation (4.35), expressed in terms of θ, becomes

$$\frac{D\theta}{Dt} = \frac{Q}{c_p}\left(\frac{p}{p_0}\right)^{-\kappa}. \tag{4.37}$$

References

Good introductory textbooks on fluid dynamics in general are those by Tritton (1988), Faber (1995) and Acheson (1990): the first two of these are specifically aimed at undergraduate physicists. The basics of atmospheric fluid dynamics, including a treatment of pressure coordinates, are covered well by Holton (1992). An excellent advanced textbook on atmospheric and oceanic fluid dynamics, going well beyond the scope of the present book, is that by Gill (1982). For elementary kinetic theory of gases see, e.g., Tabor (1991). The 'reciprocity' theorem of partial differentiation, equation (4.30), is given for example by Adkins (1983), Boas (1983) and Lyons (1995).

Problems

4.1. Starting from the ideal gas law in the form (2.6), estimate the number of molecules n/V of air per unit volume at ground level. Hence show that the mean separation between molecules there is a few nanometres. Estimate also the mean free path, given that it is of order $V/(\pi d^2 n)$, where d is an effective molecular diameter. Comment on the relevance of these estimates for the assumption that atmospheric flow may be described as the motion of a continuous fluid. At what altitude might this assumption start to break down?

4.2. Starting from the continuity equation in Eulerian form,

$$\partial\rho/\partial t + \nabla\cdot(\rho\boldsymbol{u}) = 0,$$

derive the Lagrangian form,

$$\frac{D\rho}{Dt} + \rho\nabla\cdot\boldsymbol{u} = 0,$$

by vector manipulations. Give a physical interpretation of the latter equation in terms of the rate of change of volume of a small moving 'blob' of fluid.

4.3. Consider planar motion. Suppose that the coordinates of a particle are (x, y) in an inertial frame F and (x', y') in a frame F' that rotates at a constant angular velocity Ω with respect to F. Show that

$$x = x'\cos(\Omega t) - y'\sin(\Omega t), \qquad y = y'\cos(\Omega t) + x'\sin(\Omega t),$$

if $x = x'$ and $y = y'$ at $t = 0$. Hence find (x', y') as functions of x, y and t.

As viewed by an observer in F, a particle is moving along the y-axis with constant speed v, passing through the origin at $t = 0$. Show that its path, as viewed by an observer in F', is given parametrically by $x' = vt\sin(\Omega t)$, $y' = vt\cos(\Omega t)$; show that this path is approximately a parabola near the origin and check that the motion in this region is consistent with the effect of a Coriolis force in F'.

4.4. If a ball is thrown a horizontal distance of 100 m at 30° latitude in 4 s, what is its sideways deflection due to the Coriolis force?

4.5. Estimate the Rossby numbers for the following flows and comment on your results.

(*a*) A hurricane at 20° N, with windspeed about 50 m s^{-1} and horizontal length scale about 100 km.

(*b*) A tornado in the American Midwest, with windspeed about 100 m s^{-1} and horizontal length scale about 100 m.

(*c*) Flow in a bathtub vortex.

4.6. Consider an atmospheric vortex in which the flow is steady, horizontal and independent of height, with circular streamlines, in a frame rotating with angular velocity Ω about the vertical. Neglecting friction, show that

$$\frac{u^2}{r} + 2\Omega u = \frac{1}{\rho}\frac{dp}{dr},$$

where r is the radial distance from the centre of the vortex, $u(r)$ is the (tangential) wind speed, $p(r)$ is the pressure and ρ is the density (assumed constant).

Hence explain why highs are regions of weak pressure gradients and gentle winds, but lows can have large pressure gradients and strong winds. Is the geostrophic wind an

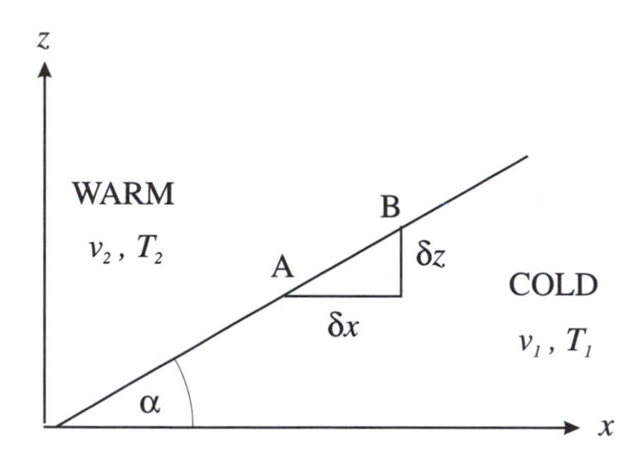

Figure 4.17 A simple model of a front.

overestimate or underestimate of the wind in a low-pressure system?

4.7. Atmospheric fronts are narrow regions of large horizontal temperature gradient. A simple model takes the front to be a sloping surface across which the temperature and along-front wind are discontinuous (with warm air overlying cold), but the pressure and cross-front wind are continuous. Take the y-axis along the front and the x-axis pointing towards the cold air and apply the hydrostatic and geostrophic wind relationships to the region AB in Figure 4.17. Hence show that the slope α at any level is related to the temperatures T_1 and T_2 and along-front winds v_1 and v_2 at that level by

$$(T_2 - T_1)g \tan\alpha = f(v_1 T_2 - v_2 T_1).$$

If $T_2 - T_1 = 3\,\mathrm{K}$ and $v_1 - v_2 = 10\,\mathrm{m\,s^{-1}}$, estimate α at 50 °N latitude.

4.8. Starting with the geostrophic and hydrostatic equations in pressure coordinates, derive the thermal windshear relationship in the form

$$\frac{\partial}{\partial p}(u, v) = \frac{R}{fp}\left(\frac{\partial T}{\partial y}, -\frac{\partial T}{\partial x}\right),$$

making clear which variables are held constant in each partial derivative. How does this equation differ from the z-coordinate version?

Introduce the variable $Z = \ln(p_0/p)$, where p_0 is a constant reference pressure, and derive the thermal wind relationship using Z as a vertical coordinate.

The temperatures in the following table are derived from satellite measurements for latitudes 50° N and 40° N and longitude 0° E. Given that the eastward wind at (45° N, 0° E) and a pressure of 10 hPa is $25\,\mathrm{m\,s^{-1}}$, estimate the eastward wind

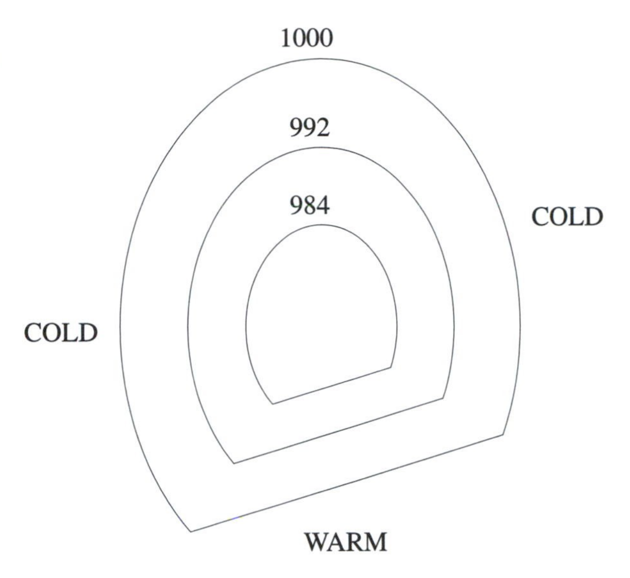

Figure 4.18 A schematic weather map of a developing cyclone.

at (45° N, 0° E) at a pressure of 1 hPa.

Pressure (hPa)	10	1
Temperature at 50° N (K)	217	252
Temperature at 40° N (K)	224	261

(Use the trapezoidal rule $\int_a^b F(Z)\,dZ \approx \frac{1}{2}[F(a)+F(b)](b-a)$.)

4.9. Figure 4.18 is a schematic 'weather map' of a developing cyclone (in the Northern Hemisphere) showing contours (isobars) of surface pressure in hPa and the regions of cold and warm air at the surface (the latter is called the 'warm sector'). Giving physical justification, mark the directions of the geostrophic winds by arrows and the positions of the cold (▲▲▲) and warm (●●●) fronts. (Note: warm air follows cold in a warm front and vice versa in a cold front.) Check that the changes in wind direction at the fronts agree with the results of Problem 4.7.

4.10. Figure 4.19 is part of a North Atlantic weather chart, for the sector 10° W–35° W and 35° N–55° N, showing surface pressure (dashed contours, in hPa) and the height of the 500 hPa surface (solid contours, in dekametres; 1 dekametre = 10 m). Show that, near 1000 hPa, the pressure drops by about 8 hPa for every 60 m of vertical ascent (take the density of air at the surface to be about 1.3 kg m^{-3}). Hence show that the surface isobars can be roughly re-interpreted as isopleths of the 1000 hPa height, and re-label them accordingly on the chart (in dekametres). Draw in contours of the thickness of the

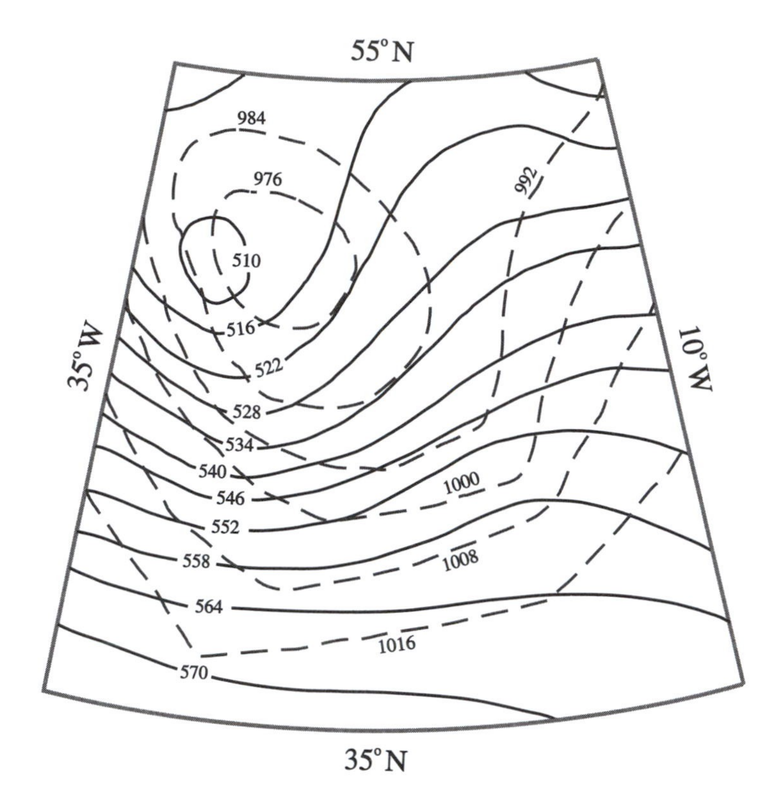

Figure 4.19 Part of a North Atlantic weather chart.

1000 hPa–500 hPa layer and shade the region where the mean temperature of this layer is largest. What is the name given to this region? What is its mean temperature?

Shade in the region of strongest geostrophic winds at 500 hPa and estimate the maximum geostrophic wind speed there.

CHAPTER FIVE

Further atmospheric fluid dynamics

In this chapter we build on the foundations laid in Chapter 4 by considering some more advanced atmospheric fluid-dynamical concepts and studying simple models of observed dynamical phenomena in the atmosphere. In Section 5.1 we introduce the important notions of *vorticity* and *potential vorticity*. In Sections 5.2 and 5.3 we make some further simplifications to the basic equations of motion, which allow us to set up simple models of two types of atmospheric *wave*: the small-scale gravity wave (in Section 5.4) and the large-scale Rossby wave (in Section 5.5). As mentioned in Chapter 1, waves are very important atmospheric phenomena, not only because they are a common feature of observations, but also because they allow one part of the atmosphere to 'communicate' with other, perhaps distant, parts of the atmosphere. In Section 5.6 we look at atmospheric *boundary layers*, regions near the Earth's surface where frictional effects become important, and in Section 5.7 we briefly touch on the important topic of atmospheric *instability*.

5.1 Vorticity and potential vorticity

The vorticity $\boldsymbol{\omega}$ is defined as the curl of the velocity vector $\boldsymbol{u} = (u, v, w)$:

$$\boldsymbol{\omega} = \nabla \times \boldsymbol{u} = \left(\frac{\partial w}{\partial y} - \frac{\partial v}{\partial z}, \quad \frac{\partial u}{\partial z} - \frac{\partial w}{\partial x}, \quad \frac{\partial v}{\partial x} - \frac{\partial u}{\partial y} \right).$$

In the special case of two-dimensional flow parallel to the x, y plane and independent of z, so that $\boldsymbol{u} = (u(x,y), v(x,y), 0)$, the vorticity has only a z component:

$$\boldsymbol{\omega} = (0, 0, \xi), \qquad \text{where} \qquad \xi = \frac{\partial v}{\partial x} - \frac{\partial u}{\partial y}. \tag{5.1}$$

The vorticity is a measure of the *local* (not *global*) rotation or

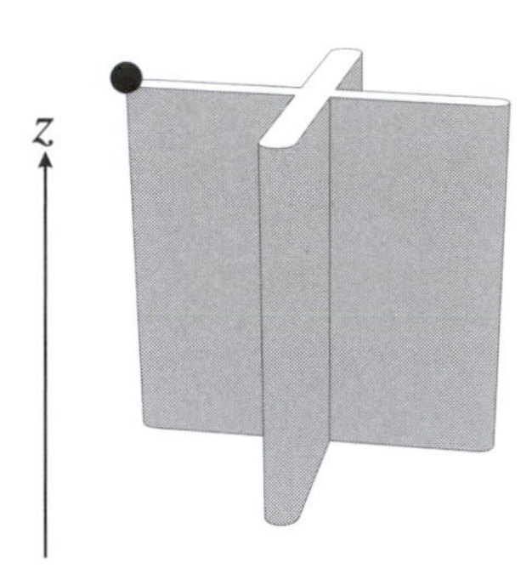

Figure 5.1 A hypothetical 'vorticity meter'. Adapted after Acheson (1990).

'spin' of the flow†. This fact is best illustrated by some simple examples:

(a) Consider two-dimensional circular flow, using plane polar coordinates r and φ, as in Section 4.8.2, so that $\boldsymbol{u} = V(r)\boldsymbol{i}_\varphi$, where $\boldsymbol{i}_\varphi$ is the unit vector in the azimuthal direction. It can be shown (see Problem 5.1) that the vorticity is in the z direction, as in equation (5.1), with

$$\xi = \frac{dV}{dr} + \frac{V}{r}. \tag{5.2}$$

Consider two special cases.

- 'Solid-body rotation', in which the fluid rotates with the same angular velocity Ω at all points, so that $V = \Omega r$. In this case equation (5.2) implies that $\xi = 2\Omega =$ constant.
- The 'point vortex', in which $V \propto r^{-1}$; in this case $\xi = 0$. (The origin $r = 0$ is a singular point and must be excluded.) This is an example of a flow that we can clearly regard as rotating in a global sense, but which has *zero* vorticity. Thus vorticity is not a signature of global rotation.

(b) Now consider two-dimensional rectilinear shear flow in Cartesian coordinates: $\boldsymbol{u} = (u(y), 0, 0)$. In this case $\xi = -du/dy$ and is generally non-zero. Such a flow may be regarded as 'non-rotating' in a global sense, but has non-zero vorticity. Again vorticity is not directly linked with global rotation.

Vorticity as a local quantity may usefully be pictured in the following way. Imagine a tiny hypothetical 'vorticity meter', consisting of four perpendicular vanes (see Figure 5.1), which is placed in a two-dimensional flow with its axis in the z direction. The vanes tend to be carried with the local flow; one vane is marked with a black dot, as shown. Close examination shows that the angular velocity of the meter is equal to *half* the local value of the vorticity.

† The word 'spin' is used in the sense of classical, not quantum, mechanics here.

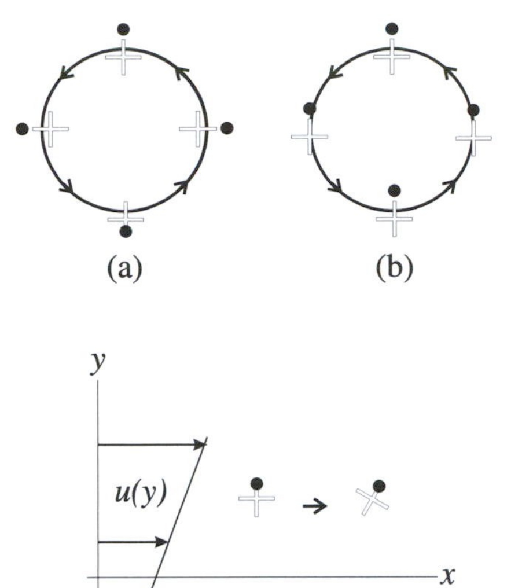

Figure 5.2 The behaviour of a hypothetical vorticity meter in (a) solid-body rotation, (b) a point vortex and (c) rectilinear shear flow. Adapted after Acheson (1990).

When it is placed in the solid-body rotation, the meter revolves in a circular orbit, as shown in Figure 5.2(a). Moreover, the azimuthal flow near the outer vane is faster than that near the inner vane, by just such an amount as to give the vane a single rotation per orbit. The meter spins with angular velocity Ω, equal to half the vorticity ξ.

When it is placed in the point vortex, however, the meter remains aligned with its original orientation as it orbits; see Figure 5.2(b). In this case the flow near the outer vane is less than that near the inner vane, by just such an amount as to keep the meter aligned. The meter has zero spin, which is consistent with the zero vorticity of the flow.

In the case of the rectilinear shear flow (Figure 5.2(c)), with $u = By$, where B is a positive constant, the flow near the upper vane in the diagram is faster than that near the lower vane. The meter spins in a clockwise sense, with negative angular velocity, which is consistent with the vorticity ξ being $-B$ here.

The concept of vorticity may be extended to rotating frames. Putting $\boldsymbol{A} = \boldsymbol{r}$ in equation (4.18), we obtain the relationship between the velocities in an inertial frame and a rotating frame:

$$\boldsymbol{u}_{\mathrm{I}} = \boldsymbol{u}_{\mathrm{R}} + \boldsymbol{\Omega} \times \boldsymbol{r},$$

where $\boldsymbol{u}_{\mathrm{I}}$ is the velocity in the inertial frame and $\boldsymbol{u}_{\mathrm{R}}$ is the velocity in the rotating frame. Taking the curl and using the vector identity

$$\nabla \times (\boldsymbol{a} \times \boldsymbol{b}) = (\boldsymbol{b} \cdot \nabla)\boldsymbol{a} - (\boldsymbol{a} \cdot \nabla)\boldsymbol{b} + \boldsymbol{a}(\nabla \cdot \boldsymbol{b}) - \boldsymbol{b}(\nabla \cdot \boldsymbol{a}), \tag{5.3}$$

where $\boldsymbol{a}$ and $\boldsymbol{b}$ are vector functions of position, we obtain

$$\boldsymbol{\omega}_{\mathrm{I}} = \boldsymbol{\omega}_{\mathrm{R}} + 2\boldsymbol{\Omega}. \tag{5.4}$$

Thus the vorticity vector in the inertial frame (the *absolute vorticity*) is that in the rotating frame (the *relative vorticity*) plus twice the rotation vector. This is consistent with the two-dimensional solid-body rotation result above, where we can regard the flow as being at rest in a frame rotating with angular speed Ω.

The concept of *potential vorticity* (PV) is more complex and will only be touched on here. The potential vorticity (sometimes called the *Rossby–Ertel potential vorticity*) is defined as

$$P = \frac{\boldsymbol{\omega}_{\mathrm{I}} \cdot \nabla\theta}{\rho}, \tag{5.5}$$

where θ is the potential temperature and ρ the fluid density. Given the Navier–Stokes equation (4.20), the mass-continuity equation (4.8) and the thermodynamic equation (4.37), it can be shown that the potential vorticity satisfies an equation of the form

$$\frac{DP}{Dt} = \text{terms involving friction and diabatic heating.}$$

A corollary of this result is that, in the absence of friction and diabatic heating, the right-hand side vanishes and the potential vorticity is *materially conserved*: in other words, each fluid blob retains a fixed value of potential vorticity as it moves around and the potential vorticity is a *tracer* of fluid motion. Together with other properties, this makes the potential vorticity a very useful fluid-flow quantity, which has been used in numerous studies as a diagnostic of atmospheric motion.

5.2 The Boussinesq approximation

In Chapter 4 we derived the following set of equations for atmospheric flow, valid on a sphere, f-plane or β-plane: the momentum equations

$$\frac{Du}{Dt} - fv + \frac{1}{\rho}\frac{\partial p}{\partial x} = F^{(x)}, \tag{5.6a}$$

$$\frac{Dv}{Dt} + fu + \frac{1}{\rho}\frac{\partial p}{\partial y} = F^{(y)}, \tag{5.6b}$$

$$\frac{Dw}{Dt} + \frac{1}{\rho}\frac{\partial p}{\partial z} + g = 0, \tag{5.6c}$$

the mass-continuity equation

$$\frac{D\rho}{Dt} + \rho\nabla\cdot\boldsymbol{u} = 0, \tag{5.6d}$$

the ideal gas law

$$p = RT\rho \tag{5.6e}$$

and the thermodynamic energy equation

$$c_p \frac{DT}{Dt} - \frac{1}{\rho}\frac{Dp}{Dt} = Q. \tag{5.6f}$$

In equations (5.6a) and (5.6b) $f = 2\Omega \sin\phi$ if spherical coordinates are used, $f = f_0$ on an f-plane and $f = f_0 + \beta y$ on a β-plane.

These equations are still quite complicated, so we now introduce the *Boussinesq approximation*, which further simplifies the mathematics while retaining much of the important physics. It is motivated by the fact that compressibility effects in the atmosphere may be neglected for many purposes, except when deep layers (greater than the density scale height, over which the density falls by a factor of e, in depth) are under consideration. In particular, sound waves may be ignored for meteorological purposes. Taking the atmosphere to be incompressible implies that equation (5.6d) decouples into two equations,

$$\nabla \cdot \boldsymbol{u} = 0, \tag{5.7}$$

$$\frac{D\rho}{Dt} = 0. \tag{5.8}$$

Equation (5.8) states that, for incompressible flow, the density is constant on following a moving fluid blob. However, this does not imply that the density is uniform everywhere; we must still allow for vertical density stratification. It is convenient to separate the density into a 'background' part $\bar{\rho}$ that depends only on height z and a deviation ρ':

$$\rho(x, y, z, t) = \bar{\rho}(z) + \rho'(x, y, z, t) \tag{5.9}$$

with a similar separation for the pressure p:

$$p(x, y, z, t) = \bar{p}(z) + p'(x, y, z, t).$$

We now approximate the vertical momentum equation (5.6c) by hydrostatic balance, as in equation (4.25):

$$\frac{\partial p}{\partial z} = -g\rho. \tag{5.10}$$

If we assume that the background state itself satisfies hydrostatic balance,

$$\frac{d\bar{p}}{dz} = -g\bar{\rho},$$

then equation (5.10) implies that the deviation also satisfies hydrostatic balance:

$$\frac{\partial p'}{\partial z} = -g\rho'. \tag{5.11}$$

Substitution of equation (5.9) into the density equation (5.8) leads to

$$\frac{D\rho'}{Dt} + w\,\frac{d\bar{\rho}}{dz} = 0. \tag{5.12}$$

This states that the density deviation ρ' of a blob changes as the blob moves up or down in the background density gradient $d\bar{\rho}/dz$. By analogy with equation (2.31) we introduce the quantity

$$N_{\mathrm{B}}^2 = -\frac{g}{\rho_0}\frac{d\bar{\rho}}{dz}, \tag{5.13}$$

where $\rho_0 = \bar{\rho}(0)$, say, a reference value of the background density. $N_{\mathrm{B}}(z)$ can be regarded as a buoyancy frequency for a stratified, incompressible fluid. Then equation (5.12) can be written

$$g\frac{D\rho'}{Dt} - \rho_0 N_{\mathrm{B}}^2 w = 0. \tag{5.14}$$

The *Boussinesq approximation* is now implemented by ignoring density variations *except* where they are coupled with gravity; that is, we replace ρ by the constant value ρ_0 in the horizontal momentum equations (5.6a) and (5.6b) but retain the full density variation (and in particular the deviation ρ') in equations (5.11) and (5.14). We therefore obtain the following *Boussinesq equations*†:

$$\frac{Du}{Dt} - fv + \frac{1}{\rho_0}\frac{\partial p}{\partial x} = F^{(x)}, \tag{5.15a}$$

$$\frac{Dv}{Dt} + fu + \frac{1}{\rho_0}\frac{\partial p}{\partial y} = F^{(y)}, \tag{5.15b}$$

$$\frac{\partial u}{\partial x} + \frac{\partial v}{\partial y} + \frac{\partial w}{\partial z} = 0, \tag{5.15c}$$

$$\frac{D}{Dt}\left(-\frac{g\rho'}{\rho_0}\right) + N_{\mathrm{B}}^2 w = 0, \tag{5.15d}$$

$$\frac{\partial p'}{\partial z} = -g\rho'. \tag{5.15e}$$

It should be noted that, under the Boussinesq approximation, the thermodynamics becomes decoupled from the dynamics. If equations (5.15) are solved, however, we can obtain the temperature using the ideal gas law

$$T = \frac{\bar{p} + p'}{R(\bar{\rho} + \rho')}.$$

A 'mass source' term is sometimes added to the right-hand side

† These equations are very similar in form to the full equations in pressure coordinates: see Section 4.9.

of equation (5.15d), by analogy with the diabatic heating Q that appears in the thermodynamic energy equation.

5.2.1 Linearised equations and energetics

Because of the presence of the $(\boldsymbol{u} \cdot \nabla)$ term in the material derivative D/Dt, equations (5.15) are nonlinear and therefore difficult to solve analytically. Nonlinear equations of this type are routinely solved on computers, for example in weather forecasting; however, numerical solutions are often difficult to interpret physically. We therefore now make the further approximation of *linearising* these equations, so that solutions can be found and analysed.

We assume that velocities and density deviations are 'small', in the sense that terms that are quadratic in these quantities and their derivatives can be neglected. For example, equation (5.15d) can be expanded as

$$-\frac{g}{\rho_0}\left(\frac{\partial \rho'}{\partial t} + \underline{u\frac{\partial \rho'}{\partial x}} + \underline{v\frac{\partial \rho'}{\partial y}} + \underline{w\frac{\partial \rho'}{\partial z}}\right) + N_{\mathrm{B}}^2 w = 0,$$

where the quadratic terms are underlined. If these are dropped we get

$$-\frac{g}{\rho_0}\left(\frac{\partial \rho'}{\partial t}\right) + N_{\mathrm{B}}^2 w = 0,$$

which is *linear* in ρ' and w. The full set of Boussinesq equations, linearised in this way and with friction neglected, is

$$u_t - fv + \frac{1}{\rho_0}p'_x = 0, \tag{5.16a}$$

$$v_t + fu + \frac{1}{\rho_0}p'_y = 0, \tag{5.16b}$$

$$u_x + v_y + w_z = 0, \tag{5.16c}$$

$$-\frac{g}{\rho_0}\rho'_t + N_{\mathrm{B}}^2 w = 0, \tag{5.16d}$$

$$p'_z + g\rho' = 0. \tag{5.16e}$$

(The incompressibility condition (5.16c) and hydrostatic equation (5.16e) were already linear and needed no further approximation here.) Note that we have introduced the useful shorthand notation of denoting partial derivatives with respect to x, y, z and t by subscripts.

An important result can be obtained by multiplying equation (5.16a) by $\rho_0 u$, equation (5.16b) by $\rho_0 v$ and equation (5.16d) by

$-g\rho'/N_B^2$ and then adding the results. A short calculation gives the *energy equation*

$$\frac{\partial}{\partial t}\frac{1}{2}\rho_0\left[u^2+v^2+\left(\frac{g\rho'}{\rho_0 N_B}\right)^2\right]+\nabla\cdot(\boldsymbol{u}p')=0. \tag{5.17}$$

In this equation, $\rho_0(u^2+v^2)/2$ is clearly the kinetic energy per unit volume of the horizontal motion, while the term involving $(\rho')^2$ can be identified as the available potential energy (with respect to a reference state at rest, with density $\bar{\rho}$ and pressure $\bar{p}$: see Section 2.6). The term $\boldsymbol{u}p'$ can be interpreted as an energy flux (analogous to the Poynting vector in electromagnetism). Overall, the equation states that the energy (kinetic plus available potential) within a volume increases if there is an energy flux into the volume and decreases if there is an energy flux out of the volume. Similar, but more complicated, versions of equation (5.17) apply for the nonlinear Boussinesq equations (5.15) and the nonlinear compressible equations (5.6).

5.3 Quasi-geostrophic motion

In Section 4.8 we derived the geostrophic approximations to equations (5.6a) and (5.6b). Under the Boussinesq approximation, neglecting variations of f, geostrophic balance can be expressed by

$$u\approx u_g=-\frac{\partial\psi}{\partial y},\qquad v\approx v_g=\frac{\partial\psi}{\partial x}, \tag{5.18}$$

where ψ is the *geostrophic streamfunction*, defined by

$$\psi\equiv\frac{p'}{f_0\rho_0}. \tag{5.19}$$

(See equations (4.26a), (4.26b) and note that we can use p' rather than p in equation (5.19) since the horizontal derivatives of $\bar{p}(z)$ vanish.) The velocity vector $(u_g, v_g, 0)$ is called the *geostrophic flow*. From equations (5.18)

$$\frac{\partial u_g}{\partial x}+\frac{\partial v_g}{\partial y}=0,$$

so by comparison with the incompressibility condition (5.15c) we can see that any vertical velocity w_g associated with the geostrophic flow must be independent of z. It is therefore zero everywhere if it vanishes at one level, say, at the ground $z=0$.

Note that, using the geostrophic streamfunction ψ, hydrostatic balance (5.15e) can be written as

$$\rho'=-\frac{f_0\rho_0}{g}\frac{\partial\psi}{\partial z}. \tag{5.20}$$

We now seek a better approximation to the nonlinear Boussinesq equations (5.15) than that given by pure geostrophic balance. One way to do this is to introduce the *ageostrophic velocity*, the difference between the true velocity and the geostrophic flow, with components

$$u_a \equiv u - u_g, \qquad v_a \equiv v - v_g, \qquad w_a \equiv w.$$

It can then be shown that the next approximation beyond geostrophic balance is given, on a β-plane and neglecting friction, by the *quasi-geostrophic equations*

$$D_g u_g - f_0 v_a - \beta y v_g = 0, \tag{5.21a}$$

$$D_g v_g + f_0 u_a + \beta y u_g = 0, \tag{5.21b}$$

$$\frac{\partial u_a}{\partial x} + \frac{\partial v_a}{\partial y} + \frac{\partial w_a}{\partial z} = 0, \tag{5.21c}$$

$$D_g\left(-\frac{g\rho'}{\rho_0}\right) + N_B^2 w_a = 0. \tag{5.21d}$$

Here

$$D_g \equiv \frac{\partial}{\partial t} + u_g\frac{\partial}{\partial x} + v_g\frac{\partial}{\partial y}$$

is the time derivative following the geostrophic flow. The quasi-geostrophic equations hold in general for large-scale, low-frequency motions, except in low latitudes; in particular they require that the Rossby number (see equation (4.27)) should be small:

$$\text{Ro} \equiv \frac{U}{fL} \ll 1, \tag{5.22}$$

and that time scales should be large compared with $1/f_0$ (a few hours, except in low latitudes). They provide a useful model for investigating many types of large-scale motion that are observed in the atmosphere.

Equations (5.21) can conveniently be combined as follows. First take $\partial/\partial x$(5.21b) – $\partial/\partial y$(5.21a) and use the mass-continuity equation (5.21c), to obtain (after careful manipulation!) the *vorticity equation*

$$D_g\zeta = f_0\frac{\partial w_a}{\partial z}, \tag{5.23}$$

where

$$\zeta \equiv f_0 + \beta y - \frac{\partial u_g}{\partial y} + \frac{\partial v_g}{\partial x} = f_0 + \beta y + \frac{\partial^2\psi}{\partial x^2} + \frac{\partial^2\psi}{\partial y^2}$$

is the z component of the *absolute vorticity* associated with the geostrophic flow; cf. equation (5.4).

The term on the right-hand side of equation (5.23) is called the

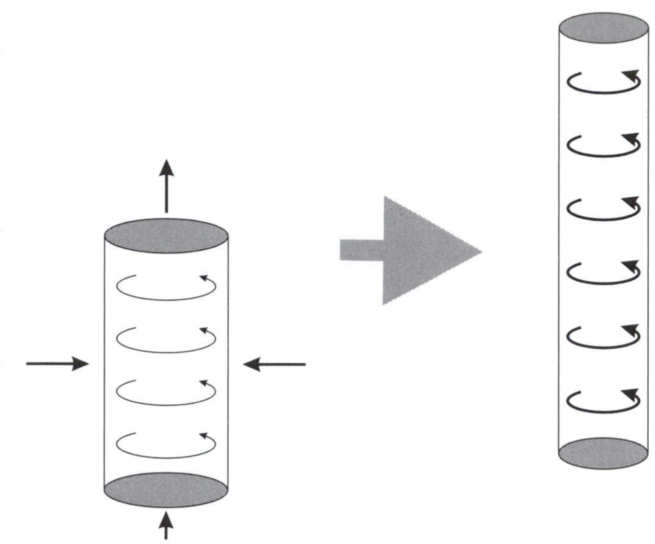

Figure 5.3 Illustrating the 'stretching' mechanism of absolute vorticity generation. The cylindrical blob of fluid on the left experiences stretching in the vertical owing to the differential vertical velocities (indicated by the vertical arrows), and shrinking in the horizontal associated with convergent horizontal velocities (indicated by the horizontal arrows). Its vertical absolute vorticity is indicated by the curved arrows. It is deformed by the velocity field into the shape shown on the right, and at the same time its vertical absolute vorticity increases.

stretching term, since it can generate vorticity by differential vertical motion. Consider a cylindrical blob of air, moving with the geostrophic flow: see Figure 5.3. If the blob enters a region where the vertical velocity w_a increases with height it is stretched vertically. Since its mass must be conserved it shrinks in the horizontal direction†, consistent with there being a horizontal convergence of velocity; see equation (5.21c). Equation (5.23) states that the z component of the absolute vorticity ζ of the blob must increase (in the Northern Hemisphere, where $f_0 > 0$) by this stretching mechanism. Conversely a cylindrical blob that is squashed in the vertical direction will suffer a decrease in ζ.

Now note that, using equations (5.21d) and (5.20), the vertical velocity can be expressed as

$$w_a = D_g\left(\frac{g\rho'}{\rho_0 N_B^2}\right) = -D_g\left(\frac{f_0}{N_B^2}\frac{\partial\psi}{\partial z}\right); \tag{5.24}$$

substitution of this expression into the vorticity equation (5.23) and further careful manipulation lead finally to the *quasi-geostrophic potential vorticity equation*

$$D_g q = 0, \tag{5.25}$$

where

$$q \equiv \zeta + \frac{\partial}{\partial z}\left(\frac{f_0^2}{N_B^2}\frac{\partial\psi}{\partial z}\right) = f_0 + \beta y + \frac{\partial^2\psi}{\partial x^2} + \frac{\partial^2\psi}{\partial y^2} + \frac{\partial}{\partial z}\left(\frac{f_0^2}{N_B^2}\frac{\partial\psi}{\partial z}\right) \tag{5.26}$$

is the *quasi-geostrophic potential vorticity* (QGPV). When friction

† Remember that, under the Boussinesq approximation, the fluid is incompressible.

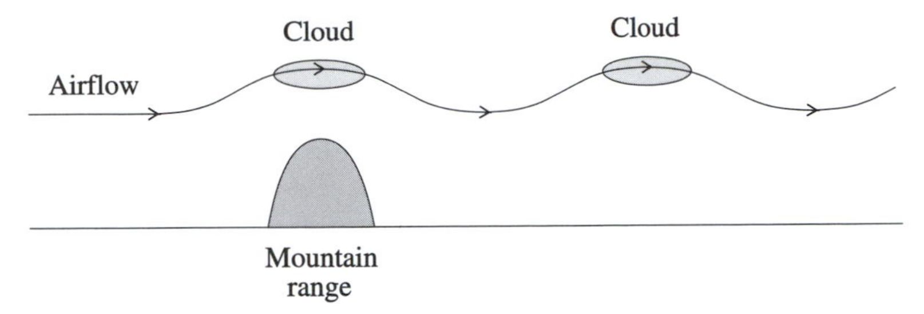

Figure 5.4 A schematic cross-section through a lee wave (or mountain wave): an internal gravity wave that may form when a stratified airflow blows over a mountain range. The wavy arrow indicates one particular streamline of the flow. The wave motion appears over and downstream of the mountain and may be made visible by clouds that form as water vapour in the moving air parcels condenses in the rising parts of the flow. The clouds disperse as the air parcels descend and the water evaporates again. Note that the wave pattern and clouds are stationary with respect to the mountain, but that the air flow blows through them.

and diabatic heating (or mass sources) are neglected, as here, the QGPV is constant following the geostrophic flow. We can 'forecast' the future behaviour of the QGPV by integrating equation (5.25) forwards in time if the current values of the QGPV and the geostrophic flow are known at all points. Moreover, the QGPV carries information about the geostrophic flow and the density, since the elliptic operator (analogous to ∇^2) on the right-hand side of equation (5.26) can be inverted, given suitable boundary conditions, to give ψ and hence p', u_g, v_g and ρ', using equations (5.19), (5.18) and (5.20).

The QGPV is somewhat analogous to the Rossby–Ertel PV (5.5), which is, however, constant following the *total* flow $\boldsymbol{u}$ in the absence of friction and diabatic heating.

5.4 Gravity waves

It was noted in Chapter 1 that wave-like motions are frequently observed in the atmosphere. One such wave is the *gravity wave*: examples of gravity waves include the lee waves that are often manifested as parallel bands of cloud downstream of mountain ranges; see Figure 5.4. Another example, derived from ground-based radar measurements and demonstrating waviness in time, is given in Figure 1.7.

The linear Boussinesq equations (5.16) are a suitable set from which to develop models of gravity waves. As in modelling other wave motions in physics (e.g., electromagnetic waves as solutions of Maxwell's equations), it is simplest to look for *plane-wave solutions* of equations (5.16). First, however, we restrict our attention to motions of comparatively small horizontal scale (for example, $\lesssim$ 100 km) so that the rotation of the Earth has a negligible effect. This allows us to simplify the equations by neglecting the Coriolis terms (i.e., by putting $f = 0$) in equations (5.16a) and (5.16b). We also assume that N_B is independent of z: this is a fairly reasonable assumption both for the troposphere and for the stratosphere.

We seek plane waves, propagating in the x, z plane, of the form

$$\left\{u, v, w, p', \rho'\right\} = \mathrm{Re}\left\{\hat{u}, \hat{v}, \hat{w}, \hat{p}, \hat{\rho}\right\} \exp[i(kx + mz - \omega t)], \tag{5.27}$$

where $\hat{u}$, etc. are complex amplitudes. Substitution of expressions (5.27) into the linear partial differential equations (5.16) then yields the algebraic equations

$$-i\omega\hat{u} + ik\hat{p}/\rho_0 = 0, \tag{5.28a}$$

$$-i\omega\hat{v} = 0, \tag{5.28b}$$

$$ik\hat{u} + im\hat{w} = 0, \tag{5.28c}$$

$$i\omega g\hat{\rho}/\rho_0 + N_{\mathrm{B}}^2\hat{w} = 0, \tag{5.28d}$$

$$im\hat{p} + g\hat{\rho} = 0. \tag{5.28e}$$

It is straightforward to eliminate, say, $\hat{u}$, $\hat{v}$, $\hat{w}$ and $\hat{\rho}$ in favour of $\hat{p}$; arbitrarily choosing $\hat{p}$ to be real, we then get

$$p' = \hat{p}\cos(kx + mz - \omega t), \tag{5.29a}$$

$$u = \frac{k\hat{p}}{\rho_0\omega}\cos(kx + mz - \omega t), \tag{5.29b}$$

$$v = 0, \tag{5.29c}$$

$$w = -\frac{k^2\hat{p}}{\rho_0\omega m}\cos(kx + mz - \omega t), \tag{5.29d}$$

$$\rho' = \frac{m\hat{p}}{g}\sin(kx + mz - \omega t). \tag{5.29e}$$

These are called the *polarisation relations* for the waves. Note that u, w and p' are in phase (or 180° out of phase) with each other and 90° out of phase with ρ'. The condition for the algebraic equations (5.28) to possess a unique solution gives the *dispersion relation*, relating the angular frequency ω to the wave-vector components k and m, for *internal gravity waves*:

$$\omega^2 = \frac{N_{\mathrm{B}}^2 k^2}{m^2}. \tag{5.30}$$

On taking the square root we obtain the two possible solutions

$$\omega = \pm\frac{N_{\mathrm{B}}k}{m}. \tag{5.31}$$

To understand the physical difference between these two solutions, we must introduce the vector group velocity

$$\boldsymbol{c}_{\mathrm{g}} = \left(c_{\mathrm{g}}^{(x)}, 0, c_{\mathrm{g}}^{(z)}\right) = \left(\frac{\partial\omega}{\partial k}, 0, \frac{\partial\omega}{\partial m}\right); \tag{5.32}$$

in particular the vertical component of the group velocity, $c_{\mathrm{g}}^{(z)}$, is

$$c_{\mathrm{g}}^{(z)} = \frac{\partial}{\partial m}\left(\pm\frac{N_{\mathrm{B}}k}{m}\right) = \mp\frac{N_{\mathrm{B}}k}{m^2}. \tag{5.33}$$

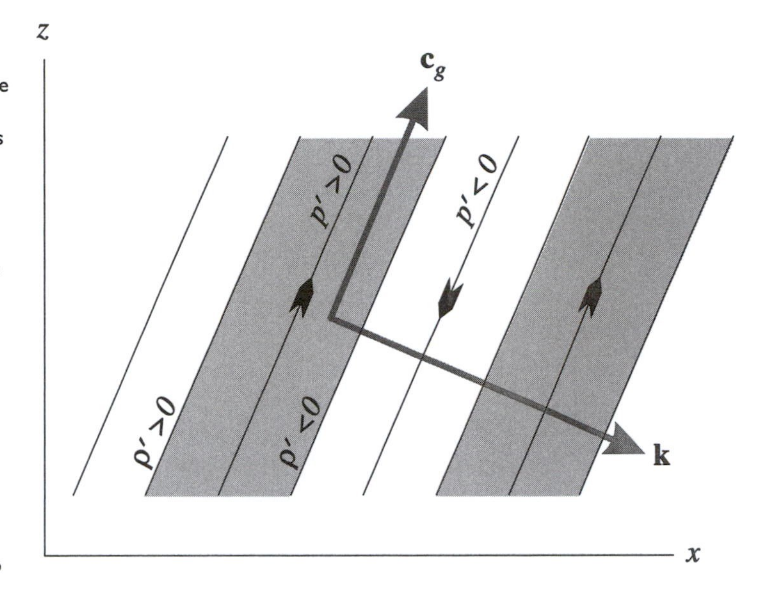

Figure 5.5 A vertical cross-section through a plane internal gravity wave with $k > 0$, $\omega > 0$ and $m < 0$ so that $c_g^{(z)} > 0$, as in the text. The thin sloping lines represent surfaces of constant phase, separated perpendicularly from one another by a quarter-wavelength. The phase surfaces on which the density and pressure disturbances take their greatest positive and negative values are marked. The phase surfaces in which the velocity vectors (which are themselves parallel to the phase surfaces) have their greatest upward and downward components are marked by long black arrowheads in the appropriate directions. The regions of upward motion are shaded. The thick grey arrow sloping obliquely upwards indicates the direction of the group velocity vector c_g and the thick grey arrow sloping obliquely downwards indicates the direction of the wave vector $\boldsymbol{k} = (k, 0, m)$. The phase surfaces move perpendicular to themselves, in the direction of $\boldsymbol{k}$, as time progresses.

We now adopt the convention that k is positive[†]. For an atmospheric internal gravity wave generated near the ground and propagating information *upwards*, we must have $c_g^{(z)} > 0$, so the *lower* sign must be chosen both in equation (5.33) and in equation (5.31), which then becomes

$$\omega = -\frac{N_B k}{m}.$$

If also $\omega > 0$, we have $m < 0$ and the phase relations between the velocity, density and pressure disturbances due to the waves may be summarised as in Figure 5.5.

Several features should be noted here.

- For this choice of signs, the phase surfaces $kx + mz - \omega t =$ constant move obliquely downwards in time, in the direction of the wave-vector $\boldsymbol{k} = (k, 0, m)$. However, the propagation of information, represented by the group velocity vector $\boldsymbol{c}_g$, is obliquely upwards.
- The velocity vector $(u, 0, w)$ is parallel to the slanting phase surfaces: fluid blobs (or parcels) oscillate up and down these surfaces.
- The precise phase relations between the velocity, density and pressure disturbances may be verified from the polarisation relations (5.29).

The hydrostatic assumption is equivalent to neglecting a term

[†] The following analysis may be repeated with the convention $k < 0$; certain changes of sign must be made in the analysis, but the physical content of the resulting equations is unchanged.

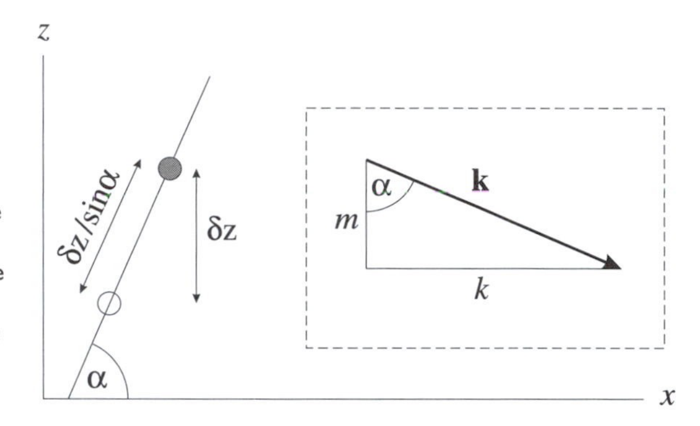

Figure 5.6 An illustration of the motion of a fluid blob in a plane internal gravity wave. The sloping line on the left of the diagram indicates a sloping phase surface. The open circle indicates the equilibrium position of the blob and the solid circle its displaced position. The inset triangle on the right of the diagram shows the wave-vector $\boldsymbol{k} = (k, 0, m)$, which is perpendicular to the phase surface.

$\rho_0 \, \partial w/\partial t$ compared with $g\rho'$, say, in the linear equation (5.16e). From the polarisation relations (5.29d) and (5.29e) it can be seen that this is valid if $k^2 \ll m^2$, i.e., if vertical wavelengths are much less than horizontal wavelengths. Equation (5.30) then shows that $\omega^2 \ll N_{\mathrm{B}}^2$ under this condition, so the hydrostatic internal gravity waves we have been considering must have angular frequencies much less than the buoyancy frequency N_{B}. Even for non-hydrostatic waves it turns out that $\omega^2 \leq N_{\mathrm{B}}^2$; typical values of the minimum period $2\pi/N_{\mathrm{B}}$ are 8 min for the troposphere and 5 min for the stratosphere.

A useful connection can be made here with the analysis of Section 2.5, in which the buoyancy frequency N in a compressible, stably stratified fluid was shown to be the frequency at which a parcel (or blob) of fluid oscillates vertically up and down. We have just seen that, in an internal gravity wave, fluid blobs do not move purely vertically, but are constrained to move along the sloping phase surfaces. These surfaces make an angle α with the horizontal, where $\sin^2\alpha = k^2/(k^2+m^2)$; see Figure 5.6. Consider a blob of unit mass: under the Boussinesq approximation its volume V remains constant and its density is approximately ρ_0, so that $\rho_0 V \approx 1$. If this blob moves a vertical distance δz upwards from its equilibrium position and hence a distance $\delta z/\sin\alpha$ along the slope, then its density is $\rho_0 N_{\mathrm{B}}^2 \delta z/g$ greater than that of its surroundings, by equation (5.13). It therefore experiences a downward buoyancy force $N_{\mathrm{B}}^2 \delta z$, which has a component $-N_{\mathrm{B}}^2 \sin\alpha \, \delta z$ *up* the sloping phase surface. Its acceleration up the slope is

$$\frac{d^2}{dt^2}\left(\frac{\delta z}{\sin\alpha}\right),$$

so, by Newton's Second Law,

$$\frac{d^2}{dt^2}\left(\frac{\delta z}{\sin\alpha}\right) + N_{\mathrm{B}}^2 \sin\alpha \, \delta z = 0$$

(cf. equation (2.30)). This implies an oscillation whose angular frequency ω is given by

$$\omega^2 = N_B^2 \sin^2 \alpha = \frac{N_B^2 k^2}{k^2 + m^2} \approx \frac{N_B^2 k^2}{m^2}$$

(the approximation holding for hydrostatic waves), in agreement with the dispersion relation (5.30).

In this section we have concentrated on internal gravity waves, the class of gravity waves whose horizontal scales are so small that the Earth's rotation can be neglected. A similar study can be carried out for waves of somewhat larger scale (horizontal wavelengths of hundreds of kilometres) and somewhat lower frequency (periods of several hours), which do feel the Earth's rotation; these are called *inertia–gravity waves*. (The waves in Figure 1.7 are of this type.) An example is given in Problem 5.4.

5.5 Rossby waves

A further class of atmospheric wave is observed on horizontal scales of thousands of kilometres and with periods of several days. These are known as *Rossby waves* or *planetary waves*: an example is shown in Figure 1.8. These are even more difficult to identify unambiguously than gravity waves, since a purely sinusoidal Rossby-wave structure can exist only in very simple background flows. However, they are conceptually very important for our understanding of many large-scale atmospheric phenomena.

Since atmospheric Rossby waves occur on large horizontal wavelengths and with low frequencies, they have low Rossby numbers (see equations (4.27) and (5.22)). It is therefore reasonable to try to model them using the quasi-geostrophic equations developed in Section 5.3. We consider small-amplitude disturbances to a uniform zonal background flow $(U, 0, 0)$, where U is a constant; by equation (5.18) this uniform flow corresponds to a geostrophic streamfunction $\Psi = -Uy$. For the total flow (the background plus a small disturbance) we therefore take

$$\psi = -Uy + \psi',$$

substitute into the QGPV equation (5.25) and neglect terms that are quadratic in ψ'.

The QGPV for this flow is

$$q = f_0 + \beta y + \mathcal{L}\psi', \tag{5.34}$$

where $\mathcal{L}$ is the elliptic operator

$$\mathcal{L} = \frac{\partial^2}{\partial x^2} + \frac{\partial^2}{\partial y^2} + \frac{\partial}{\partial z}\left(\frac{f_0^2}{N_B^2}\frac{\partial}{\partial z}\right),$$

and the QGPV equation (5.25) linearises to

$$\left(\frac{\partial}{\partial t} + U\frac{\partial}{\partial x}\right)\mathcal{L}\psi' + \beta\frac{\partial \psi'}{\partial x} = 0. \tag{5.35}$$

We take N_B to be constant, for simplicity, and look for plane-wave solutions to this equation as in Section 5.4 (but now allowing variations in y as well) by substituting

$$\psi' = \mathrm{Re}\,\hat{\psi}\exp[i(kx + ly + mz - \omega t)],$$

where $\hat{\psi}$ is a complex amplitude, and obtain the dispersion relation for Rossby waves:

$$\omega = kU - \frac{\beta k}{k^2 + l^2 + f_0^2 m^2/N_B^2}. \tag{5.36}$$

In simple modelling it is usual for k, l and ω to be specified in advance, so that the vertical wavenumber m can then be found from equation (5.36):

$$m = \pm\frac{N_B}{f_0}\left(\frac{\beta}{U - (\omega/k)} - \left(k^2 + l^2\right)\right)^{1/2}. \tag{5.37}$$

Several features are immediately evident from the dispersion relation (5.36):

- The β-effect is crucial to the existence of these waves: putting $\beta = 0$ in equation (5.36) gives $\omega = kU$, which corresponds to 'waves' that are merely carried along with the background flow U.
- Equation (4.24) shows that $\beta > 0$. The zonal phase speed of the waves,

  $$c \equiv \frac{\omega}{k} = U - \frac{\beta}{k^2 + l^2 + f_0^2 m^2/N_B^2},$$

  therefore always satisfies

  $$U - c > 0; \tag{5.38}$$

  that is, the wave crests and troughs (which move with the phase speed) move *westward* with respect to the background flow.
- For given real values of the horizontal wavenumbers k and l, we get vertical propagation if m is real and non-zero (and vertical evanescence if m is imaginary). Vertical propagation

therefore corresponds to $m^2 > 0$ and this implies that

$$U - c = \frac{\beta}{k^2 + l^2 + f_0^2 m^2/N_B^2} < U_c \equiv \frac{\beta}{k^2 + l^2}, \qquad (5.39)$$

where U_c depends on the horizontal wavelengths of the wave. Putting equations (5.38) and (5.39) together, we therefore find that for vertical propagation we must have

$$0 < U - c < U_c.$$

In particular, for *stationary* waves, whose crests and troughs do not move with respect to the ground (which will be true for waves forced by the background flow moving over continent-scale topography or stationary heat sources) so that $c = 0$, we obtain the *Charney–Drazin criterion*

$$0 < U < U_c. \qquad (5.40)$$

This states that stationary waves propagate vertically only in eastward background flows ($U > 0$) that are not too strong ($U < U_c$); moreover, since U_c increases with increasing horizontal wavelength, 'long waves' propagate vertically under a wider range of eastward flows than do 'short waves'. This is consistent with observations in the Northern Hemisphere winter stratosphere, where winds are eastward and stationary Rossby waves have large horizontal scales, and with observations of the summer stratosphere, where winds are westward and stationary Rossby waves are absent.

- Rossby waves with large horizontal scales have small values of $k^2 + l^2$ and their zonal phase speed approaches

$$c_{\text{long}} = U - \frac{\beta N_B^2}{f_0^2 m^2},$$

whereas Rossby waves with short horizontal scales have large values of $k^2 + l^2$ and their zonal phase speed approaches the speed U of the background flow. This strong variation of phase speed with wavelength implies that the waves are strongly *dispersive*: an initial disturbance composed of a number of different wavelengths will tend to break up, or disperse, in time, as the various wavelength components propagate away at different phase speeds.

- We can define a group velocity vector

$$\boldsymbol{c}_g = \left(c_g^{(x)}, c_g^{(y)}, c_g^{(z)}\right) = \left(\frac{\partial \omega}{\partial k}, \frac{\partial \omega}{\partial l}, \frac{\partial \omega}{\partial m}\right),$$

as for gravity waves (equation (5.32)), except that here ω is taken to depend on l as well as on k and m, so $\boldsymbol{c}_g$ will have a non-zero y component†. In particular, we find

$$c_{\mathrm{g}}^{(z)} = \frac{\partial \omega}{\partial m} = \frac{2 f_0^2 \beta k m}{N_{\mathrm{B}}^2 \left(k^2 + l^2 + \dfrac{f_0^2 m^2}{N_{\mathrm{B}}^2} \right)^2};$$

if we choose $k > 0$ by convention, we see that the vertical component of the group velocity is positive and that the waves propagate information upwards, if $m > 0$: this determines the choice of sign in equation (5.37). For upward-propagating waves the phase surfaces $kx + ly + mz - \omega t =$ constant slope westward with height: this slope is observed for Rossby waves in the stratosphere.

A further possibility occurs when the waves are independent of height: this is equivalent to taking $m = 0$ above. Suppose also that there is no background flow, so that $U = 0$; then equation (5.36) reduces to

$$\omega = -\frac{\beta k}{k^2 + l^2}$$

and the phase speed $c = \omega / k$ is westward, which is consistent with equation (5.38). This result can also be obtained by a physical argument, based on vorticity, as follows.

When there is no z dependence the disturbance part of the QGPV, $\mathcal{L}\psi'$ in equation (5.34), reduces to the disturbance vorticity,

$$\xi' = \frac{\partial^2 \psi'}{\partial x^2} + \frac{\partial^2 \psi'}{\partial y^2};$$

cf. equation (5.26). The total QGPV is then $q = f_0 + \beta y + \xi'$, of which $f_0 + \beta y$ is the background contribution due to the Earth's rotation and spherical geometry, represented by the β-effect. However, from equation (5.25) q is conserved following fluid blobs (assuming that they move essentially with the geostrophic flow). Therefore a northward-moving blob, which encounters an increasing $f_0 + \beta y$, must lose some of its disturbance vorticity ξ', whereas a southward-moving blob must gain some ξ'.

Consider a line of blobs, labelled A, B, C, etc., initially lying along a line of latitude $y = y_0$; see Figure 5.7. Suppose that these blobs are displaced into the sinusoidal pattern indicated by the solid wavy line: blob A moves southwards, so its value of ξ' increases, as indicated by the anticlockwise arrow in the figure. This induces an anticlockwise rotation in the local velocity field and, in particular, blob B is encouraged to move further south. The value of ξ' associated with blob B itself increases, inducing

† For stationary waves we must put $c = \omega = 0$ *after* differentiation. Note that these waves can still propagate information, even though their phase surfaces do not move: this is another consequence of their dispersive nature.

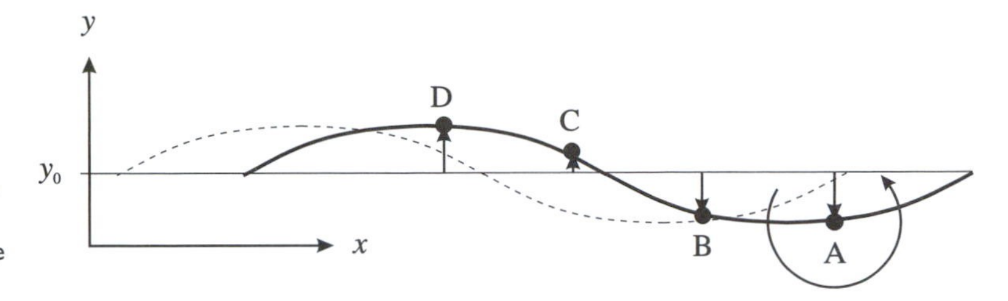

Figure 5.7 Illustrating the Rossby-wave mechanism, in terms of conservation of potential vorticity by moving fluid blobs in the case in which the waves are independent of height. See the text for details.

anticlockwise rotation near B, which tends to move C southwards and A northwards again. Applying this kind of argument to each blob, we find that, after a short time, the *pattern* of the blobs has moved westwards, to the position indicated by the dashed wavy line, even though *each individual blob* only oscillates north–south. A self-sustaining, westward-moving Rossby-wave pattern emerges, as expected from the theory given above.

5.6 Boundary layers

5.6.1 General considerations

So far in this chapter we have ignored the effects of friction in the atmosphere. This is reasonable in the models of linear internal gravity waves and Rossby waves investigated in the previous sections, since friction usually has only a small influence on these waves. However, frictional effects can sometimes be very important, especially in the lowest kilometre or so of the atmosphere. This region is sometimes called the *atmospheric boundary layer* – a terminology that echoes the use of 'boundary layer' in fluid dynamics to mean a thin, frictionally controlled layer near a boundary.

In the derivation of the Navier–Stokes equation (4.17) in Chapter 4 we included the effects of molecular viscosity. However, for most atmospheric purposes, the molecular viscosity is far too small to influence the dynamics directly. On the other hand, small-scale eddies can lead to a momentum transfer that in some respects resembles molecular momentum transfer. As we shall see later, this small-scale eddy transport is sometimes represented in terms of an 'eddy viscosity' that is similar to, but much larger than, the molecular viscosity. There are serious problems with this approach, though; for example, small-scale eddies may have organised structures, quite unlike the random nature of molecules, which may even lead to an eddy viscosity that is *negative*! We shall therefore progress as far as possible without explicit use of the eddy-viscosity concept.

We consider the frictional stress τ due to small-scale processes (with scales of a few hundred metres or less) acting on the larger scales of motion in the atmosphere and assume that this stress is horizontal:

$$\boldsymbol{\tau} = \left(\tau^{(x)}, \tau^{(y)}, 0\right).$$

As in equation (4.14), the vertical gradient of $\boldsymbol{\tau}$ implies a horizontal force per unit mass on the larger scales, given by

$$F^{(x)} = \frac{1}{\rho_0}\frac{\partial \tau^{(x)}}{\partial z}, \qquad F^{(y)} = \frac{1}{\rho_0}\frac{\partial \tau^{(y)}}{\partial z}, \tag{5.41}$$

in the Boussinesq approximation. On substituting equations (5.41) into the f-plane versions of the Boussinesq horizontal momentum equations (5.15a) and (5.15b) and linearising, we obtain

$$\frac{\partial u}{\partial t} - f_0 v = -\frac{1}{\rho_0}\frac{\partial p}{\partial x} + \frac{1}{\rho_0}\frac{\partial \tau^{(x)}}{\partial z}, \tag{5.42a}$$

$$\frac{\partial v}{\partial t} + f_0 u = -\frac{1}{\rho_0}\frac{\partial p}{\partial y} + \frac{1}{\rho_0}\frac{\partial \tau^{(y)}}{\partial z}. \tag{5.42b}$$

We now write the horizontal velocity components as

$$u = u_p + u_\tau, \qquad v = v_p + v_\tau, \tag{5.43}$$

where the subscript p denotes the *pressure-driven flow*, satisfying

$$\frac{\partial u_p}{\partial t} - f_0 v_p = -\frac{1}{\rho_0}\frac{\partial p}{\partial x}, \tag{5.44a}$$

$$\frac{\partial v_p}{\partial t} + f_0 u_p = -\frac{1}{\rho_0}\frac{\partial p}{\partial y} \tag{5.44b}$$

and the subscript τ denotes the *frictional stress-driven flow*, satisfying

$$\frac{\partial u_\tau}{\partial t} - f_0 v_\tau = \frac{1}{\rho_0}\frac{\partial \tau^{(x)}}{\partial z}, \tag{5.45a}$$

$$\frac{\partial v_\tau}{\partial t} + f_0 u_\tau = \frac{1}{\rho_0}\frac{\partial \tau^{(y)}}{\partial z}. \tag{5.45b}$$

This separation can be made since equations (5.42) are linear.

We now assume that the frictional stress is important only in a boundary layer of depth D above flat ground at $z = 0$; thus $\tau^{(x)}$ and $\tau^{(y)}$ are non-zero for $0 \le z < D$ but vanish for $z \ge D$, in the 'free atmosphere'; see Figure 5.8. Integrating equations (5.45) through the depth of the boundary layer, we then get

$$\frac{\partial U_\tau}{\partial t} - f_0 V_\tau = -\frac{1}{\rho_0}\tau_0^{(x)}, \tag{5.46a}$$

$$\frac{\partial V_\tau}{\partial t} + f_0 U_\tau = -\frac{1}{\rho_0}\tau_0^{(y)}, \tag{5.46b}$$

where

$$U_\tau = \int_0^D u_\tau \, dz, \qquad V_\tau = \int_0^D v_\tau \, dz$$

are called the *Ekman volume transports*, representing the horizontal fluxes of volume within the boundary layer, and $\tau_0^{(x)}$ and $\tau_0^{(y)}$ are the surface stresses, exerted by the ground on the lowest layer of the atmosphere. If the flow is *steady*, with $\partial/\partial t = 0$, then equations (5.46) give

$$(U_\tau, V_\tau, 0) = \frac{1}{\rho_0 f_0}\left(-\tau_0^{(y)}, \tau_0^{(x)}, 0\right) = \frac{1}{\rho_0 f_0}\boldsymbol{k} \times \boldsymbol{\tau}_0, \tag{5.47}$$

where $\boldsymbol{k} = (0, 0, 1)$ is the unit vertical vector and hence show that the Ekman volume transport in the boundary layer is *perpendicular* to the surface stress. It is important to note that this result does not depend on the details of the vertical variation of the frictional stresses (other than that they vanish above the boundary layer) or on any assumption about the eddy viscosity.

Turning now to the pressure-driven flow, equations (5.44) show that, when this flow is steady, it must also be geostrophic:

$$u_p = -\frac{1}{\rho_0 f_0}\frac{\partial p}{\partial y}, \qquad v_p = \frac{1}{\rho_0 f_0}\frac{\partial p}{\partial x};$$

cf. equations (4.26a) and (4.26b). Then the incompressibility condition (5.15c) implies that the vertical velocity depends only on the stress-driven flow:

$$\frac{\partial w}{\partial z} = -\left(\frac{\partial u}{\partial x} + \frac{\partial v}{\partial y}\right) = -\left(\frac{\partial u_\tau}{\partial x} + \frac{\partial v_\tau}{\partial y}\right). \tag{5.48}$$

Integration of equation (5.48) through the depth of the boundary layer yields

$$\left[w\right]_0^D = -\left(\frac{\partial U_\tau}{\partial x} + \frac{\partial V_\tau}{\partial y}\right); \tag{5.49}$$

the term on the right-hand side of this equation represents the horizontal convergence of the Ekman volume transport. The ground is flat, so $w = 0$ at $z = 0$, and, since the fluid is assumed incompressible, the horizontal convergence in the boundary layer must be balanced by an upward flow w_{D} out of the top of the boundary layer, at $z = D$; see Figure 5.8. Conversely, horizontal divergence in the boundary layer is balanced by a downward flow into the top of the boundary layer. Using equations (5.47) and (5.49) this vertical flow can be written in terms of the surface stress component:

$$w_{\mathrm{D}} = -\frac{1}{\rho_0 f_0}\left(\frac{\partial \tau_0^{(x)}}{\partial y} - \frac{\partial \tau_0^{(y)}}{\partial x}\right). \tag{5.50}$$

The velocity w_{D} is called the *Ekman pumping* or *suction* velocity

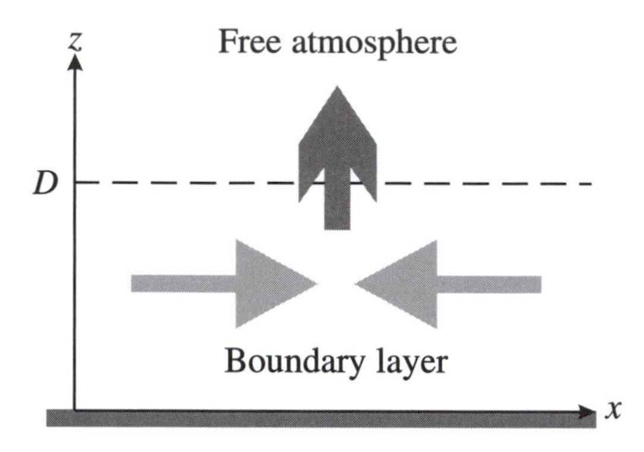

Figure 5.8 A schematic illustration of the frictional boundary layer, with the free atmosphere above. The dashed line indicates the notional top of the boundary layer. Note that horizontal convergence in the boundary layer (indicated by the lightly shaded arrows) is balanced by flow out of the top of the boundary layer (indicated by the darker arrow). This is the Ekman pumping or suction effect.

and equation (5.50) shows that it is proportional to the 'curl' of the surface stress.

5.6.2 The laminar Ekman layer

The theory so far has given information only on the vertically integrated properties of the boundary layer. To look at the vertical structure it is necessary to make an assumption about the relationship between the frictional stress τ and the velocity. A simple relationship of this kind is to assume that the stress components are proportional to the vertical gradients of the horizontal velocity:

$$\tau^{(x)} = \rho_0 \nu \frac{\partial u}{\partial z}, \qquad \tau^{(y)} = \rho_0 \nu \frac{\partial v}{\partial z}. \tag{5.51}$$

This is clearly analogous to the molecular viscous stress law (4.13). The quantity ν is the *kinematic eddy viscosity* and corresponds to the viscosity coefficient η, defined in equation (4.13), divided by the density. As noted above, we must use the eddy-viscosity concept with caution.

Using equations (5.51) we now find the vertical structure of the flow within the boundary layer, again assuming that the flow is steady. In principle the eddy viscosity ν could vary with z, but we take it to be constant for simplicity. We take $f_0 > 0$, corresponding to the Northern Hemisphere.

Under these conditions the pressure-driven flow (u_p, v_p) is again geostrophic; since the boundary layer is shallow, we take (u_p, v_p) to be independent of z within the boundary layer and equal to the large-scale flow (also assumed geostrophic) just above the boundary layer. For simplicity, we take this large-scale flow to be purely zonal, but varying with the northward distance y; thus

$$u_p = U(y) = \text{constant}, \qquad v_p = 0, \tag{5.52}$$

say. (The calculation can be re-worked with $v_p \neq 0$; see Problem 5.8.)

The boundary conditions on the total flow are

$$(u, v) \to (0, 0) \qquad \text{as} \qquad z \to 0,$$

since, with friction, there can be no flow at the ground, and

$$(u, v) \to (U, 0) \qquad \text{for} \qquad z \gg D,$$

assuming that the flow tends to the geostrophic, pressure-driven, value far above the boundary layer. In terms of the stress-driven flow, these become

$$(u_\tau, v_\tau) \to (-U, 0) \qquad \text{as} \qquad z \to 0, \tag{5.53}$$

$$(u_\tau, v_\tau) \to (0, 0) \qquad \text{for} \qquad z \gg D, \tag{5.54}$$

using equations (5.43). From equations (5.45), assuming steady flow and using the stress law (5.51), the stress-driven flow satisfies the coupled differential equations

$$-f_0 v_\tau = \nu \frac{d^2 u_\tau}{dz^2}, \qquad f_0 u_\tau = \nu \frac{d^2 v_\tau}{dz^2}. \tag{5.55}$$

These can be reduced to a single fourth-order differential equation for u_τ or v_τ, but are more easily solved by introducing the complex quantity

$$\lambda_\tau \equiv u_\tau + i v_\tau,$$

which, from equations (5.55), satisfies the second-order equation

$$\frac{d^2 \lambda_\tau}{dz^2} = \frac{i f_0}{\nu} \lambda_\tau.$$

The two solutions of this are

$$\lambda_\tau \propto \exp\left[\pm \left(\frac{i f_0}{\nu}\right)^{1/2} z\right] = \exp\left\{\pm \left[(1+i)\left(\frac{f_0}{2\nu}\right)^{1/2}\right] z\right\}. \tag{5.56}$$

To satisfy the boundary condition (5.54) we must choose the minus sign in equation (5.56), to ensure exponential decay with z. Then, taking the real and imaginary parts of λ_τ, applying the boundary condition (5.53) and adding the pressure-driven flow (5.52), we get the full solution

$$u = U[1 - e^{-z/h} \cos(z/h)], \tag{5.57a}$$

$$v = U e^{-z/h} \sin(z/h), \tag{5.57b}$$

where

$$h \equiv \left(\frac{2\nu}{f_0}\right)^{1/2}.$$

This is *Ekman's solution* and the corresponding boundary layer is called the *Ekman layer*.

We can calculate the corresponding Ekman pumping veloc-

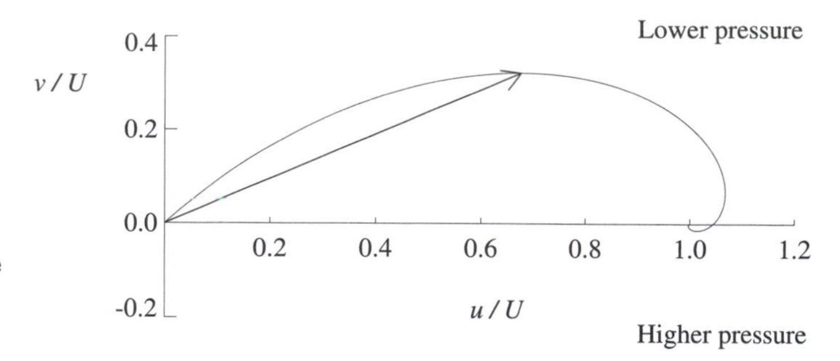

Figure 5.9 The Ekman spiral: a hodograph of horizontal wind components (normalised by U) in the Ekman layer. The horizontal axis gives the direction of the geostrophic wind above the boundary layer, so the pressure decreases along the vertical axis. The diagonal arrow indicates the wind vector at a height $z = h\pi/4$.

ity (5.50) using the stress law (5.51) and equations (5.57):

$$w_D = \left(\frac{\nu}{2f_0}\right)^{1/2}\left(-\frac{dU}{dy}\right) = \frac{1}{2}h\xi, \tag{5.58}$$

where $\xi = -dU/dy$ is the relative vorticity of the free-atmosphere flow. This relationship between w_D and ξ also holds when the free-atmosphere flow is not purely zonal. Equation (5.58) shows, for example, that the Ekman pumping velocity is upwards under a cyclone in the free atmosphere (for which $\xi > 0$) and downwards under an anticyclone (for which $\xi < 0$).

Figure 5.9 gives a graphical representation (a hodograph) of the horizontal velocity vector (u, v) as a function of z/h. This diagram is called the *Ekman spiral*; note how the velocity vector spirals in towards the free-atmosphere flow $(U, 0)$. It is conventional to define the 'depth' of the Ekman layer to be the height at which the horizontal flow first becomes parallel to the free-atmosphere flow; this corresponds to $z = \pi h = \pi(2\nu/f_0)^{1/2}$. Putting this equal to 1 km, the approximate depth of the atmospheric boundary layer, allows us to get an order-of-magnitude estimate of the value of the eddy viscosity as $\nu \sim 5\,\mathrm{m^2\,s^{-1}}$.

The Ekman spiral also shows that the deflection of the wind in the boundary layer is mostly to the low-pressure side of the geostrophic, large-scale flow. Recall from Section 4.8 that the geostrophic flow blows along the isobars; the Ekman-layer analysis indicates how this is altered in the presence of friction. However, it should be noted that the assumptions that have been made in this calculation are seldom fully satisfied in the atmosphere, owing to the presence of temporal variations and, probably, to the inappropriateness of the eddy-viscosity concept. As a result, a pure Ekman spiral is hardly ever observed in the real atmospheric boundary layer.

5.7 Instability

In Section 2.5 we noted that a compressible atmosphere is statically stable if the squared buoyancy frequency N^2 is positive and

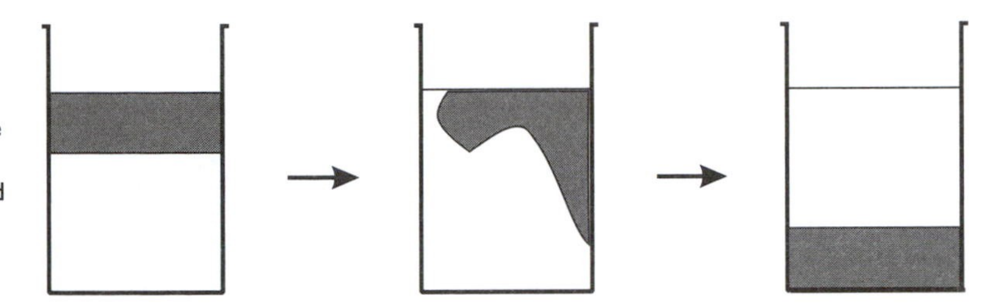

Figure 5.10 Illustrating static instability in an incompressible fluid. In the left-hand panel, dense fluid (shaded) is introduced over light fluid (unshaded). Static instability leads to a redistribution of the fluids; in the final stable state, shown in the right-hand panel, the dense fluid is entirely below the light fluid.

statically unstable if it is negative. A similar criterion, but involving N_B^2 rather than N^2, applies to an incompressible atmosphere under the Boussinesq approximation. Static instability is probably the simplest type of fluid instability to appreciate: at the most basic level it occurs if a layer of dense, incompressible fluid of density ρ_2 is introduced over a layer of lighter incompressible fluid of density $\rho_1 < \rho_2$ in a closed container (Figure 5.10). In this case the dense fluid will fall through the lighter fluid, perhaps in a complicated way, leading eventually to a statically stable state in which the dense fluid is entirely below the light fluid.

A more complex example would be a statically unstable region of finite depth in a compressible atmosphere. Here parcels displaced adiabatically upwards are 'lighter' (i.e., have a higher potential temperature θ) than their surroundings and continue to rise, whereas parcels displaced downwards are 'heavier' (i.e., have a lower θ) than their surroundings and continue to fall. Eventually a statically stable state is reached, in which low values of θ are at the bottom and higher values are at the top, so that $N^2 = (g/\theta)(d\theta/dz) > 0$.

A third case is one in which local regions of static instability are set up by heating from below, leading to *convective instability* (see Figure 5.11 for a laboratory analogue). In the presence of heating of this kind in the atmosphere we may expect much of the forced region to approach neutral stability, with N^2 approaching zero and the lapse rate approaching the dry adiabatic value Γ_a (see Section 2.5) or the saturated adiabatic value Γ_s (see Section 2.8) according to whether the air is unsaturated or saturated.

On large scales, the rotation of the Earth allows more subtle types of instability, which can lead to important atmospheric disturbances such as cyclones and anticyclones and other weather phenomena. We briefly discuss the two most important of these: *baroclinic instability* and *barotropic instability*.

5.7.1 Baroclinic instability

Consider a steady zonal flow $(U(z), 0, 0)$ in the Northern Hemisphere that increases with height and work with the Boussinesq equations on an f-plane. From geostrophic balance (5.18) and hy-

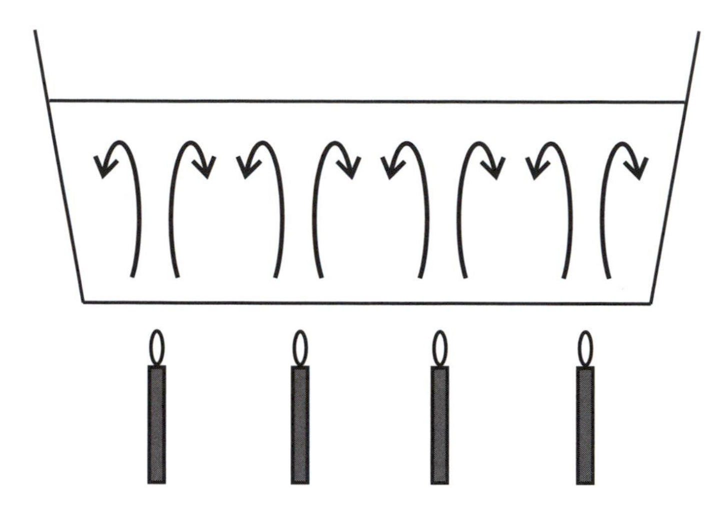

Figure 5.11 A schematic illustration of convection patterns in a fluid heated from below and cooled from above.

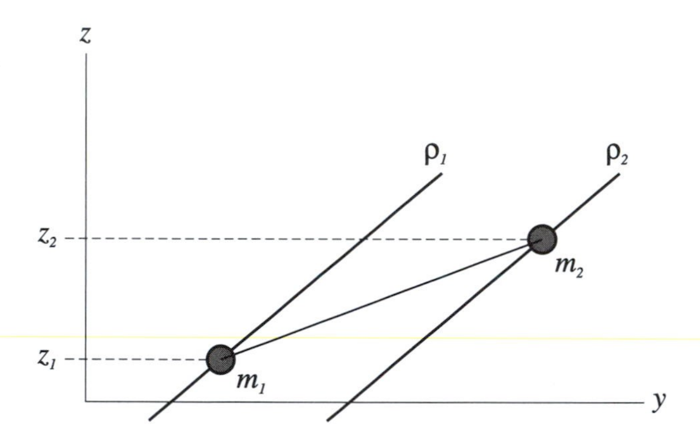

Figure 5.12 Hypothetical interchange of two parcels from surfaces of different densities.

drostatic balance (5.20) we can derive a thermal windshear equation (cf. equation (4.28b))

$$f_0 \frac{dU}{dz} = \frac{g}{\rho_0} \frac{\partial \rho}{\partial y},$$

where $\rho = \bar{\rho}(z) + \rho'$. Since $dU/dz > 0$ and $f_0 > 0$ this implies that $\partial\rho/\partial y > 0$, i.e., the density increases with northward distance. Now the slope α of the density surfaces in the y, z plane satisfies

$$\tan \alpha \equiv \left(\frac{\partial z}{\partial y}\right)_\rho = -\frac{\partial \rho/\partial y}{\partial \rho/\partial z}$$

by the reciprocity theorem of partial differentiation, equation (4.30). Assuming that the flow is statically stable, so that $\partial\rho/\partial z < 0$, it follows that $\tan\alpha > 0$. Thus a zonal wind that increases with height is associated, through thermal windshear balance, with density surfaces that slope polewards and upwards.

We now show that this configuration of density surfaces may be able to release potential energy. Suppose that air parcels, each of volume V, are somehow interchanged between two density-surfaces $\rho = \rho_1$ and $\rho = \rho_2$, where $\rho_2 > \rho_1$, as shown in Figure 5.12. That is, parcel A, of mass $m_1 = \rho_1 V$ and at initial height z_1, is swapped with parcel B, of mass $m_2 = \rho_2 V$ and initial height z_2, without any other part of the fluid being disturbed. The initial potential energy of the two parcels alone is $g(m_1 z_1 + m_2 z_2)$ while their final potential energy is $g(m_2 z_1 + m_1 z_2)$. There is therefore an increase of potential energy of the two parcels (and hence of the whole fluid, since no other changes take place) given by

$$\Delta E_{\mathrm{P}} = g(m_2 z_1 + m_1 z_2) - g(m_1 z_1 + m_2 z_2)$$

$$= -g(m_2 - m_1)(z_2 - z_1) = -gV(\rho_2 - \rho_1)(z_2 - z_1).$$

However, $\rho_2 > \rho_1$, so, if $z_2 > z_1$, as in Figure 5.12 (that is, the heavier parcel is initially higher than the lighter parcel), then $\Delta E_{\mathrm{P}} < 0$; there is thus a *decrease* of potential energy as a result of the interchange. This can occur, despite the stable density stratification, because the density surfaces are sloping more steeply than the line joining the two parcels. There is then a possibility that the potential energy released by the interchange may be converted to kinetic energy, leading to unstable motions. These motions are an example of *baroclinic instability*, also known as *sloping convection*†. Of course this kind of interchange of parcels is highly idealised, so further analysis is needed to determine whether it is ever dynamically possible.

A simple model, which is more realistic and also dynamically consistent, is *Eady's model* of baroclinic instability, in which the background flow is taken to be a linear function of height, $U = \Lambda z$, between rigid horizontal boundaries at $z = D$ and $z = -D$ on an f-plane, with N_{B} constant. The corresponding background streamfunction is $\Psi = -\Lambda yz$ and the background density is a linear function of y.

Linearised quasi-geostrophic disturbances to this flow are considered: these satisfy a linearised QGPV equation of the form

$$\left(\frac{\partial}{\partial t} + U\frac{\partial}{\partial x}\right)\left(\frac{\partial^2 \psi'}{\partial x^2} + \frac{f_0^2}{N_{\mathrm{B}}^2}\frac{\partial^2 \psi'}{\partial z^2}\right) = 0, \tag{5.59}$$

(cf. equations (5.25) and (5.35)). The boundary conditions are that the vertical velocity $w_{\mathrm{a}} = 0$ at $z = 0$ and D; from equation (5.24) this implies that $D_{\mathrm{g}}(\partial\psi/\partial z) = 0$ at the boundaries. On linearisation

† The word *baroclinic* in this context refers to the height variation of the zonal wind or the corresponding latitudinal variation of the density or temperature.

about the background flow this gives

$$\left(\frac{\partial}{\partial t} + U\frac{\partial}{\partial x}\right)\frac{\partial \psi'}{\partial z} - \Lambda\frac{\partial \psi'}{\partial x} = 0 \qquad \text{at} \qquad z = \pm D. \tag{5.60}$$

Normal mode solutions of equations (5.59) and (5.60) of the form

$$\psi' = \operatorname{Re}\hat{\psi}(z)\exp[ik(x - ct)] \tag{5.61}$$

are now sought, where $k > 0$ and the phase speed c may be complex: $c = c_r + ic_i$. It is found that $c_i > 0$ for a certain range of wavenumbers, corresponding to a normal mode that grows in time like $\exp(kc_i t)$: this exponential growth is a signal of instability. The quantity kc_i is called the *growth rate*; it is the inverse of the *e*-folding time for the instability. A brief sketch of the details is as follows.

Substitution of (5.61) into (5.59) gives

$$ik(U - c)\left(-k^2\hat{\psi} + \frac{f_0^2}{N_B^2}\hat{\psi}''\right) = 0$$

and hence, if $U(z) \neq c$,

$$\hat{\psi}'' - K^2\hat{\psi} = 0, \qquad \text{where} \qquad K = \frac{N_B k}{f_0};$$

note that this corresponds to a perturbation QGPV that is zero everywhere. The solutions can be written

$$\hat{\psi} = A\cosh(Kz) + B\sinh(Kz), \tag{5.62}$$

where A and B are complex constants. Substitution of (5.61) into the boundary conditions (5.60) gives

$$\left(z - \frac{c}{\Lambda}\right)\hat{\psi}' - \hat{\psi} = 0 \qquad \text{at} \qquad z = \pm D.$$

When expression (5.62) is substituted into each of these boundary conditions, the resulting equations added and subtracted and then A and B eliminated, we obtain the dispersion relation

$$c^2 = -\frac{\Lambda^2}{K^2}[\coth(KD) - KD][KD - \tanh(KD)]. \tag{5.63}$$

It can be shown that the right-hand side of equation (5.63) is negative for $0 < KD \lesssim 1.2$ and positive otherwise. Therefore we get growing modes ($c_i > 0$) for non-zero wavenumbers k less than about $1.2f_0/(N_B D)$ or horizontal wavelengths greater than about $5.2N_B D/f_0$. The maximum value of c_i occurs where $KD \approx 0.8$; this corresponds to a maximum growth rate kc_i of about $0.31\Lambda f_0/N_B$. For $k \gtrsim 1.2f_0/(N_B D)$, c is real and no growth occurs; see Figure 5.13.

Although the Eady model is highly simplified, it can be generalised in various ways. The unstable modes that result bear some

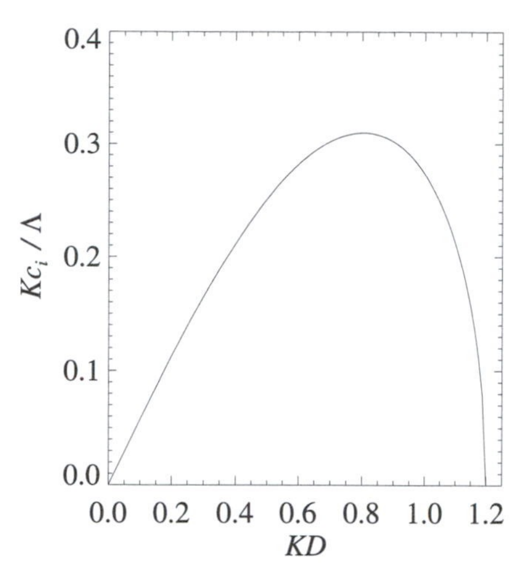

Figure 5.13 Kc_i/Λ as a function of KD for unstable normal modes in Eady's model of baroclinic instability. Note that the ordinate is proportional to the growth rate kc_i and the abscissa is proportional to the horizontal wavenumber k.

resemblance to developing cyclones and anticyclones in the midlatitude troposphere and also to certain unstable waves observed in laboratory analogues of the atmosphere (see Section 8.4).

5.7.2 Barotropic instability

A different type of instability can occur for basic winds $U(y)$ that vary with the northward distance y but not with height z; this kind of flow structure is called *barotropic*. Taking the disturbance streamfunction ψ' to vary only with y and retaining the β-effect, we get

$$\left(\frac{\partial}{\partial t} + U\frac{\partial}{\partial x}\right)\left(\frac{\partial^2\psi'}{\partial x^2} + \frac{\partial^2\psi'}{\partial y^2}\right) + (\beta - U'')\,\frac{\partial\psi'}{\partial x} = 0,$$

from the linearised QGPV equation (5.35). We seek normal mode solutions of the form

$$\psi = \mathrm{Re}\,\hat{\phi}(y)\exp[ik(x-ct)], \tag{5.64}$$

where again $k > 0$ and c may be complex, and obtain the second-order ordinary differential equation

$$(U-c)\left(\hat{\phi}'' - k^2\hat{\phi}\right) + (\beta - U'')\hat{\phi} = 0. \tag{5.65}$$

This must be solved subject to suitable boundary conditions; convenient (though not physically very realistic) ones are that the northward velocity should vanish on east–west boundaries at $y = 0, D$, say. Hence

$$\hat{\phi} = 0 \qquad \text{at} \qquad y = 0,\ D. \tag{5.66}$$

Equations (5.65) and (5.66) form a system with complex eigenvalues $c^{(n)} = c_r^{(n)} + ic_i^{(n)}$, $n = 1, 2, 3, \ldots$ and corresponding complex eigenfunctions $\hat{\phi}^{(n)}(y)$. Unlike the Eady-problem case, these equations cannot generally be solved analytically; however, they are straightforward to solve by numerical methods.

The physical interpretation of these solutions is as follows. Suppose that the zonal flow $U(y)$ is initially undisturbed, for $t < 0$. At time $t = 0$ it is given a small perturbation ψ_0; this perturbation can be Fourier-analysed into components of differing wavenumber k; moreover each of these components can in turn be resolved into a linear combination of the eigenfunctions for that value of k. Each term (or mode) in this sum then evolves in time according to equation (5.64), with its own complex phase speed. Modes for which $c_i \leq 0$ are *stable*: their amplitudes never grow significantly in time. However, any modes for which $c_i > 0$ will grow exponentially in time like $\exp(kc_i t)$; these are called *unstable modes* and the mode with the largest c_i will eventually outgrow the others and dominate the disturbance. Thus if unstable modes exist, most initial disturbances will generally excite them and the flow will rapidly be distorted. (Of course, the exponential growth in this case means that the assumption of a small disturbance amplitude will eventually break down for sufficiently large t, whereupon the model ceases to be valid.)

This *barotropic instability* mechanism is believed to be responsible for a number of types of large-scale weather disturbance observed in the tropical troposphere, as well as eddies in the stratosphere. On the other hand, if there are no unstable modes, initial disturbances will not grow and the flow is not affected very much by the perturbation imposed at $t = 0$.

A useful general result is the *Rayleigh–Kuo criterion*, which states that if $\beta - U''$ does not change sign in the region $0 \leq y \leq D$ then all modes are stable. It does *not* guarantee that unstable modes will *always* occur if $\beta - U''$ vanishes somewhere; however, the latter condition must hold if there is to be any chance of finding unstable modes.

References

The physical interpretation of vorticity given here follows the approach of Acheson (1990). Vector-calculus identities like equation (5.3) are given in many mathematics texts, such as those by Boas (1983) and Stephenson (1973). Detailed derivations of the quasi-geostrophic equations are given for example by Pedlosky (1987) and Gill (1982). The original derivation of equation (5.40) was given in the pioneering paper by Charney and Drazin (1961). For further details on potential vorticity, boundary layers and baroclinic and barotropic instability see Holton (1992) and Gill (1982). Gill in particular gives further information on the Eady model, including the spatial structure of the unstable modes, and on the Rayleigh–Kuo criterion.

Problems

5.1. Show that the two-dimensional circular flow $\boldsymbol{u} = V(r)\boldsymbol{i}_\varphi$ introduced in Section 5.1 can be written in plane Cartesian coordinates (x, y, z) as

$$\boldsymbol{u} = \frac{V}{(x^2 + y^2)^{1/2}}(-y, x, 0);$$

hence or otherwise verify that the z component of the vorticity is

$$\xi = \frac{dV}{dr} + \frac{V}{r},$$

and that the other components are zero.

5.2. Pure *inertial oscillations* can be modelled by neglecting horizontal pressure gradients in the linearised Boussinesq equations (5.16a) and (5.16b) on an f-plane:

$$u_t - f_0 v = 0, \qquad v_t + f_0 u = 0.$$

Define $\tilde{u} = u + iv$ and solve for (u, v), given that $(u, v) = (u_0, 0)$ at $t = 0$. Given that *particle displacements* (X, Y) satisfy

$$\frac{\partial X}{\partial t} = u, \qquad \frac{\partial Y}{\partial t} = v,$$

show that the particle trajectories for these oscillations are circular. Calculate the radius of the circle if the particle speed is (a) $1\,\mathrm{m\,s^{-1}}$ and (b) $10\,\mathrm{m\,s^{-1}}$.

5.3. Calculate the horizontal and vertical air-parcel displacements X and Z associated with an internal gravity wave, defined by $X_t = u$ and $Z_t = W$ (cf. Problem 5.2), given that these displacements vanish when $x = z = t = 0$. Show that the air parcels oscillate in straight lines perpendicular to the vector $(k, 0, m)$.

Find the period (in minutes) of an internal gravity wave of horizontal wavelength 100 km and vertical wavelength 5 km in the Earth's mesosphere, where $N_B^2 = 3 \times 10^{-4}\,\mathrm{s^{-2}}$. How long (in minutes) does the information associated with this wave take to propagate through a vertical distance of 20 km? If the maximum horizontal wind fluctuation (peak-to-peak) due to the wave is $2\,\mathrm{m\,s^{-1}}$, find the maximum horizontal and vertical distances traversed by an air parcel.

5.4. *Inertia–gravity waves* are the generalisation of the internal gravity waves of Section 5.4 to the case when $f_0 \neq 0$. Look for linear plane-wave solutions of the form (5.27) and show in particular that

$$\rho_0 \hat{u} = \frac{\omega k \hat{p}}{\omega^2 - f_0^2}, \qquad \rho_0 \hat{v} = \frac{-ikf_0 \hat{p}}{\omega^2 - f_0^2} \tag{5.67}$$

and that the dispersion relation is

$$\omega^2 = f_0^2 + \frac{N_B^2 k^2}{m^2}.$$

Note that the northward velocity v must be non-zero.

What is the minimum angular frequency of these waves? Show that, for a given frequency and vertical wavelength, these waves have a larger horizontal wavelength than do the corresponding internal gravity waves.

5.5. Using equations (5.67) and the hydrostatic equation, calculate the mean kinetic energy per unit volume $\overline{K}$ and the mean available potential energy per unit volume $\overline{P}$, averaged over one wave period. How does the ratio $\overline{P}/\overline{K}$ behave when $\omega \to f_0$ and when $\omega \gg f_0$? To what kind of wave does the latter limit refer? Show that equipartition between mean kinetic energy and mean available potential energy does not occur for long waves.

5.6. Starting with the Rossby-wave dispersion relation (5.36), consider waves in the absence of a background flow, with $l = 0$. For fixed m put $b = f_0 m/N_B$ and sketch ω as a function of k, for $k > 0$, with particular attention to the limits $k \to 0$ and $k \gg b$. Show that Rossby waves cannot exist if the wave period is shorter than a critical value. Estimate this critical period at 45° N for waves of vertical wavelength 10 km, given $N_B^2 = 5 \times 10^{-4}\ \mathrm{s^{-2}}$, a value representative of the stratosphere.

5.7. Consider Rossby waves that vary with x and y but are independent of depth z, in a uniform zonal flow U. Show that stationary waves can exist only if the zonal (x) wavelength is greater than a certain value, L say. Calculate L at 60° N for (a) $U = 50\ \mathrm{m\,s^{-1}}$ and (b) $U = 150\ \mathrm{m\,s^{-1}}$. Compare this with the length of a latitude circle and comment.

5.8. Take $(u, v) = (u_p + u_\tau, v_p + v_\tau)$, where $u_p(x, y)$, $v_p(x, y)$ is a geostrophic flow and $u_\tau(z)$ and $v_\tau(z)$ satisfy the steady laminar Ekman-layer equations

$$-fv_\tau = \nu \frac{\partial^2 u_\tau}{\partial z^2}, \qquad fu_\tau = \nu \frac{\partial^2 v_\tau}{\partial z^2},$$

(f is a constant, taken to be positive, for the Northern Hemisphere). The boundary conditions are $(u, v) = 0$ on $z = 0$ and $(u, v) \to (u_p, v_p)$ as $\zeta \to \infty$, where $\zeta = z/h$ and $h = (2\nu/f)^{1/2}$.

Put $\lambda_\tau = u_\tau + iv_\tau$ and show that

$$\lambda_\tau = -(u_p + iv_p)e^{-(1+i)\zeta}. \tag{5.68}$$

Hence verify that

$$u = u_p(1 - e^{-\zeta}\cos\zeta) - v_p e^{-\zeta}\sin\zeta,$$
$$v = v_p(1 - e^{-\zeta}\cos\zeta) + u_p e^{-\zeta}\sin\zeta.$$

Choose axes such that $v_p = 0$ and draw the Ekman spiral, showing how the vector (u, v) varies as ζ increases from 0 to ∞. Note that friction causes a significant cross-isobar flow in the Ekman layer: in which direction is this flow, in general? Sketch the relative directions of the pressure-gradient force, the Coriolis force and the viscous force at some point above the ground but well within the boundary layer, assuming that $u_p > 0$.

Using equation (5.68), calculate $\int_0^\infty u_\tau\, dz$ and $\int_0^\infty v_\tau\, dz$. What do these quantities represent? Show that the vertical velocity at the top of the Ekman layer is given by

$$w_\tau = \left(\frac{\nu}{2f}\right)^{1/2}\left(\frac{\partial v_p}{\partial x} - \frac{\partial u_p}{\partial y}\right).$$

If the boundary layer depth is of order 1 km and $f = 10^{-4}\,\mathrm{s}^{-1}$, estimate the order of magnitude of ν. If the geostrophic wind changes by $20\,\mathrm{m\,s^{-1}}$ over a horizontal distance of 1000 km, estimate the order of magnitude of w_τ. Give the direction and order of magnitude of the horizontal stress at the ground at a point where the geostrophic wind is $10\,\mathrm{m\,s^{-1}}$. (Assume that $\rho = 1\,\mathrm{kg\,m^{-3}}$.)

5.9. Suppose that friction is represented by linear *Rayleigh friction*

$$F^{(x)} = -ru, \qquad F^{(y)} = -rv,$$

instead of by an eddy-viscosity assumption. Use this representation in the linear Boussinesq momentum equations on an f plane to model the following situation.

During the daytime, convection is active, leading to a strong frictional coupling with the ground (modelled by taking r to be large, $r = 10^{-5}\,\mathrm{s}^{-1}$) and the atmosphere can be assumed to be in a steady state at this time. At sunset, convection is quenched and r becomes zero. Assuming that the pressure gradient is purely in the y direction, is independent of height and does not change with time, show that the subsequent motion consists of geostrophic motion, superimposed on which is an inertial oscillation of amplitude $ru_g/(r^2 + f^2)^{1/2}$, where u_g is the geostrophic velocity.

Calculate the amplitude of the oscillation at 30° N when the flow is $20\,\mathrm{m\,s^{-1}}$ from the west. Show that after 12 h the meridional flow will be the reverse of its daytime value. (Assume that night-time conditions last this long.)

5.10. Consider a region of the atmosphere, in the Northern Hemisphere, that is in geostrophic and hydrostatic balance. Make the Boussinesq approximation and assume that the wind is in the zonal direction and increases linearly with height, while the density ρ decreases linearly with height.

Show that the surfaces of constant density slope upwards towards the pole at an angle α to the horizontal given by

$$\tan\alpha = f\Lambda/N_B^2,$$

where Λ is the vertical wind shear and N_B is the buoyancy (Brunt–Väisälä) frequency.

Now suppose that two air parcels of equal volume V, densities ρ_1 and ρ_2 (where $\rho_2 > \rho_1$) and heights z_1 and z_2 are interchanged, while the rest of the atmosphere remains undisturbed. Under what condition is potential energy released by this process?

Suppose that $\delta\rho = \rho_2 - \rho_1$ and the distance δs between the parcels are small. Show that the amount of potential energy released is given by

$$\Delta P = \rho_0 V (N_B\,\delta s)^2 (\sin\phi\cos\phi\tan\alpha - \sin^2\phi),$$

where ϕ is the angle between the line joining the parcels and the horizontal. Show that ΔP reaches a maximum ΔP_{max} when $\phi = \alpha/2$. Given that α is small, so that $\alpha \approx \tan\alpha$, show that

$$\Delta P_{max} \approx \rho_0 V \left(\frac{f\Lambda\,\delta s}{2N_B}\right)^2.$$

Estimate the angle to the horizontal at which air parcels would move to maximise the release of potential energy near 60° N, given a vertical windshear of $2\,\mathrm{m\,s^{-1}\,km^{-1}}$ and a buoyancy period of 8 min.

5.11. Apply the Eady model of baroclinic instability to a region of the atmosphere near 50° N in which the mean zonal wind varies by $30\,\mathrm{m\,s^{-1}}$ over a depth of 10 km and $N_B \simeq 10^{-2}\,\mathrm{s^{-1}}$. Estimate the time (in days) taken by the fastest-growing Eady mode to grow in amplitude by a factor of e and estimate the zonal wavelength of this mode. What is the shortest zonal wavelength for which instability can occur under the given conditions?

CHAPTER SIX

Stratospheric chemistry

In keeping with the emphasis on atmospheric *physics* in this book, the purpose of the present chapter is to illustrate the use of basic physical principles in the study of some aspects of atmospheric chemistry, rather than to provide a comprehensive treatment of atmospheric chemistry as a whole. We therefore focus on stratospheric chemistry, which provides some simple yet important applications of the basic principles and also some examples of interactions between chemistry and dynamics.

In Section 6.1 we outline some of the basic thermodynamics of chemical reactions, while in Section 6.2 we introduce some elementary aspects of chemical kinetics, including the concepts of reaction rates and chemical lifetimes. In Section 6.3 we focus on bimolecular reactions and show how physical reasoning can give an expression for the reaction rate. The process of photodissociation is introduced in Section 6.4. Once these basic ideas have been established, we apply them to stratospheric ozone in Section 6.5, first describing the Chapman theory (which involves oxygen compounds only) and then introducing the effects of catalytic cycles. The principles of chemical transport by atmospheric flows are discussed in Section 6.6, with a qualitative description of the main global-scale meridional transport structures in the middle atmosphere. Finally, in Section 6.7, we bring several of these ideas together in a general description of the processes implicated in the formation of the Antarctic ozone hole.

6.1 Thermodynamics of chemical reactions

Consider a chemical reaction in which reactants A and B lead to products C and D:

$$A + B \rightarrow C + D. \quad (6.1)$$

In the laboratory this reaction takes place within a fixed reaction vessel; in the atmosphere we have to imagine it taking place within a given 'parcel' or 'blob' of air, with no transfer of mass into or out of the parcel. If the reaction takes place at constant pressure p, which is usually true for reactions in the atmosphere, then it follows from the First Law of Thermodynamics in the form (2.18) and from equation (2.19) that the heat ΔQ *supplied to* the parcel is given, assuming reversibility, by

$$\Delta Q = \Delta H.$$

Here ΔH is the *enthalpy of reaction*, that is the sum of the enthalpies of the products minus the sum of the enthalpies of the reactants. If $\Delta H > 0$, so that heat must be supplied for the reaction to proceed, the reaction is called *endothermic*, whereas if $\Delta H < 0$, so that heat is liberated, the reaction is called *exothermic*.

The *standard molar enthalpy of formation* of a compound is the enthalpy of reaction associated with the formation of one mole of the compound from its constituent elements at a standard temperature (e.g., 25° C = 298.15 K) and pressure (1 atm). This is often denoted by $\Delta H_f^\ominus$, the superscript $\ominus$ referring to the standard temperature and pressure. Tables of $\Delta H_f^\ominus$ for a variety of atmospheric gases are available in atmospheric chemistry texts. (The corrections required to convert the enthalpy of reaction to other atmospheric temperatures and pressures are fairly small.)

Applying these ideas to reaction (6.1) we obtain, at standard temperature and pressure (STP), the molar enthalpy of reaction

$$\Delta H^\ominus = \Delta H_f^\ominus(\mathrm{C}) + \Delta H_f^\ominus(\mathrm{D}) - \Delta H_f^\ominus(\mathrm{A}) - \Delta H_f^\ominus(\mathrm{B}). \tag{6.2}$$

Given the enthalpies of formation on the right-hand side of this equation, we can determine whether the reaction is endothermic or exothermic. As an example, consider the recombination of oxygen (O) atoms to molecular oxygen (O_2) near the mesopause,

$$\mathrm{O} + \mathrm{O} + \mathrm{M} \rightarrow \mathrm{O_2} + \mathrm{M}. \tag{6.3}$$

Here M is an arbitrary air molecule, which is required to satisfy conservation of energy and momentum in the reaction. Typically M is either of the most abundant atmospheric molecules, N_2 or O_2. Reaction (6.3) is found to have a negative enthalpy of reaction $\Delta H \approx -500\,\mathrm{kJ\,mol^{-1}}$, and is thus exothermic; it therefore contributes to the diabatic heating of the atmosphere; see Section 4.10.

Conversely, the photolysis of O_2 to form two O atoms has $\Delta H \approx +500\,\mathrm{kJ\,mol^{-1}}$ (the same enthalpies of formation apply as those in reaction (6.3)), so this reaction is endothermic: some source of energy is required to drive it. One possibility, which actually occurs in the atmosphere, is that this energy is supplied by solar photons. A simple calculation (see Problem 6.1) shows that the relevant photons must have wavelengths of less than about 240 nm. The standard notation for such a reaction is

$$O_2 + h\nu \rightarrow 2O, \tag{6.4}$$

$h\nu$ here referring to the photon energy required for photolysis.

Thermodynamics also allows us to predict the *direction* in which a reaction will proceed; for example, whether $A + B \rightarrow C + D$ or $C + D \rightarrow A + B$. Consider a natural (or spontaneous) change for a 'system' (here taken to be our parcel of reacting gases) immersed in a 'heat bath' (the surrounding atmosphere) at fixed temperature and pressure. The Second Law of Thermodynamics implies that the Gibbs free energy $G = U + pV - TS$ (see Section 2.10) of the system (excluding the surroundings) must *decrease* during this change†; i.e.,

$$\Delta G \leq 0. \tag{6.5}$$

An equation analogous to equation (6.2) relates the overall *molar Gibbs free energy of reaction* at standard temperature and pressure, $\Delta G^{\ominus}$, to the molar Gibbs free energies of formation of the gases taking part in a reaction. Tabulations of the latter then allow us to determine whether $\Delta G^{\ominus}$ is negative (so that the reaction can occur spontaneously at STP), positive (so that it cannot) or zero (so that there is equilibrium between the gases involved).

Since the enthalpy $H = U + pV$ and $T = T_0$ is constant, we have the following relation among the molar Gibbs free energy of reaction ΔG, the molar enthalpy of reaction ΔH and the change in molar entropy of the system ΔS:

$$\Delta G = \Delta H - T_0 \Delta S.$$

6.2 Chemical kinetics

We now consider how *rapidly* chemical reactions occur, starting with the simplest *first-order* or *unimolecular* reaction, in which one molecule of a single reactant A breaks down to one molecule of product B and one molecule of product C:

$$A \rightarrow B + C.$$

Representing the number density of A as [A] molecules per unit volume, we define the *reaction rate* R_A as the rate of decrease of

† The proof starts with the statement that the entropy of the system plus the surroundings cannot decrease during the change. Consideration of the heat and work interactions between the system and its surroundings and use of the First Law and the facts that $T = T_0$ and $p = p_0$ are constant during the change then lead to (6.5). Details are given in standard books on thermodynamics.

[A] or equivalently the rate of increase of [B] or [C]:

$$R_A = -\frac{\partial[A]}{\partial t} = \frac{\partial[B]}{\partial t} = \frac{\partial[C]}{\partial t} \equiv k_A[A]. \tag{6.6}$$

The term on the far right-hand side of this equation defines the *rate coefficient* k_A, which has units of s^{-1}; its inverse, the *chemical lifetime* τ_A of A satisfies

$$\tau_A = \frac{1}{k_A} = \frac{[A]}{\left|\frac{\partial[A]}{\partial t}\right|}.$$

In practice the rate coefficient is independent of [A] and t, so we can integrate equation (6.6) to get

$$[A] = [A]_0 e^{-k_A t} = [A]_0 e^{-t/\tau_A},$$

where the subscript 0 indicates the value at time $t = 0$. This shows that the number density of A decays exponentially in time, decreasing by a factor e in one chemical lifetime.

Next we consider *second-order* or *bimolecular* reactions, of the type considered in equation (6.1):

$$A + B \rightarrow C + D.$$

In this case the reaction rate R_A and the rate coefficient k_{AB} satisfy

$$R_A = -\frac{\partial[A]}{\partial t} = -\frac{\partial[B]}{\partial t} = \frac{\partial[C]}{\partial t} = \frac{\partial[D]}{\partial t} \equiv k_{AB}[A][B]. \tag{6.7}$$

We can define an instantaneous lifetime for A for this reaction by

$$\tau_A = \frac{[A]}{\left|\frac{\partial[A]}{\partial t}\right|} = \frac{1}{k_{AB}[B]}.$$

In this case we only get a solution of the form

$$[A] = [A]_0 e^{-t/\tau_A} \tag{6.8}$$

if $[B] \gg [A]$ so that [B] can be considered to be approximately constant in time; see Problem 6.3. Third-order (termolecular) and higher-order reactions can also be considered; some examples are discussed later.

We now consider several simultaneous reactions. First suppose that A has three loss processes, but no production, described by

$$A \rightarrow \text{products} \qquad (k_1), \tag{6.9a}$$

$$A + B \rightarrow \text{products} \qquad (k_2), \tag{6.9b}$$

$$A + C + D \rightarrow \text{products} \qquad (k_3), \tag{6.9c}$$

where the k_i are the rate coefficients. The reaction rate R_A includes contributions from each of these processes. It is given by

$$R_A = -\frac{\partial[A]}{\partial t} = k_1[A] + k_2[A][B] + k_3[A][C][D].$$

The corresponding lifetime for A is

$$\tau_A = \frac{[A]}{\left|\frac{\partial[A]}{\partial t}\right|} = \frac{1}{k_1 + k_2[B] + k_3[C][D]}$$

and [A] decays exponentially with time if [B], [C] and [D] are approximately constant. (The lifetime may still give an instantaneous measure of the fractional rate of change of [A] even if [B], [C] and [D] are varying.)

Now suppose that, in addition to the loss processes described by reactions (6.9), there is also production of A, given by

$$E + F \rightarrow A + G \qquad (k_4);$$

in this case the reaction rate is

$$R_A = -\frac{\partial[A]}{\partial t} = k_1[A] + k_2[A][B] + k_3[A][C][D] - k_4[E][F].$$

This shows that a *steady state* is possible, in which $\partial[A]/\partial t = 0$, with a balance between production and loss of A and an equilibrium number density of A given by

$$[A] = \frac{k_4[E][F]}{k_1 + k_2[B] + k_3[C][D]}.$$

It should be emphasised that the equations considered in this section apply to a fixed *volume* of reacting chemicals. They need modification when applied to a moving parcel of air, whose volume may be changing with time; see Section 6.6.

6.3 Bimolecular reactions

Some physical insight into the nature of chemical reactions can be gained by considering a special case, that of bimolecular reactions, which we now represent by

$$A + BC \rightarrow (ABC) \rightarrow AB + C. \tag{6.10}$$

This starts with initial molecules A and BC; BC breaks into fragments B and C and then A, B and C pass through an intermediate 'transient reaction complex' ABC, before forming the final molecules AB and C. A schematic diagram of the intermediate complex is given in Figure 6.1: initially the distance r_{BC} between B and C is small and the distance r_{AB} between A and B is large. During the reaction, the situation is reversed, r_{AB} becoming small

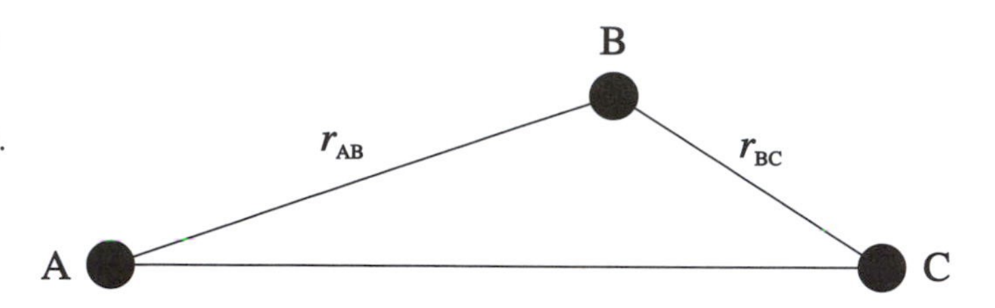

Figure 6.1 A schematic sketch of the transient reaction complex (ABC) involved in reaction (6.10).

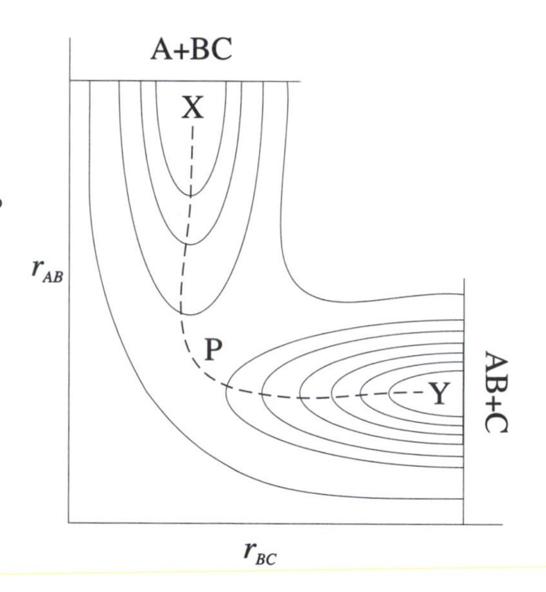

Figure 6.2 A contour plot of potential energy for reaction (6.10), showing the path of the reaction as a dashed line that proceeds from X, over the saddle point P, to Y.

and r_{BC} becoming large. In the special case in which the angle $\angle$ABC is fixed, the process can be represented in a two-dimensional contour plot of the potential energy as a function of r_{AB} and r_{BC}; see Figure 6.2. This shows that, if sufficient energy is available, ABC can proceed up the 'potential energy valley' X, associated with A + BC, over the 'saddle point' or 'col' at P and down into the 'valley' Y associated with AB + C, following the dashed line in Figure 6.2.

This idea is further illustrated in Figure 6.3, a plot of the potential energy against a 'reaction coordinate' r, which indicates how far the reaction has proceeded. This diagram is a vertical slice, along the dashed line, through the potential energy surface whose contours are given in Figure 6.2. Reaction (6.10) proceeds from left to right in Figure 6.3: an *activation energy* $E_1 = H_{ABC} - H_{A+BC}$ is required to get from the valley X over the saddle point at P. In the case shown, a larger amount of energy $E_2 = H_{ABC} - H_{AB+C}$ is regained on the further side, in valley Y, so the reaction is exothermic, with an enthalpy of reaction $\Delta H = E_1 - E_2 < 0$. On the other hand, the reverse reaction AB + C $\rightarrow$ A + BC, proceeding from right

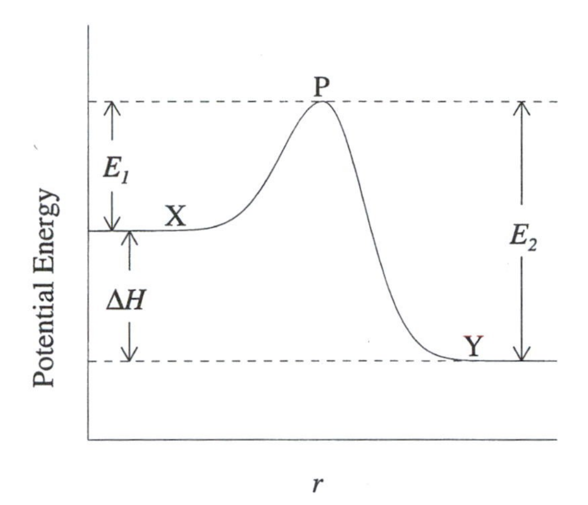

Figure 6.3 A plot of the potential energy for reaction (6.10), as a function of the reaction coordinate r.

to left in Figure 6.3, is endothermic, with an enthalpy of reaction $\Delta H = E_2 - E_1 > 0$.

A crude *collision theory* provides a simple model for the temperature dependence of the bimolecular rate coefficient. The kinetic theory of gases shows that the number of collisions, per unit volume per unit time, between molecules A and BC is $[\mathrm{A}][\mathrm{BC}]\sigma\bar{c}$, where σ is the collision cross-section

$$\sigma = \pi(r_\mathrm{A} + r_\mathrm{BC})^2,$$

r_A and r_BC being the notional radii of the molecules (taken to be hard spheres in this model). The molecules are taken to have masses m_A and m_BC, respectively, and $\bar{c}$ is the mean relative speed of the molecules at temperature T:

$$\bar{c} = \left(\frac{8kT}{\pi m_\mathrm{r}}\right)^{1/2},$$

where m_r is the reduced mass $m_\mathrm{A} m_\mathrm{BC}/(m_\mathrm{A} + m_\mathrm{BC})$ of the molecules. However, not all collisions lead to reaction: the colliding molecules must possess sufficient kinetic energy to overcome the energy barrier E_1; the probability of this occurring is given by the Boltzmann factor $\exp[-E_1/(kT)]$, where k is Boltzmann's constant. The mean rate of loss of molecules A per unit volume is therefore

$$-\frac{\partial[\mathrm{A}]}{\partial t} = [\mathrm{A}][\mathrm{BC}]\pi(r_\mathrm{A} + r_\mathrm{BC})^2 \left(\frac{8kT}{\pi m_\mathrm{r}}\right)^{1/2} e^{-E_1/(kT)} \equiv k_2[\mathrm{A}][\mathrm{BC}],$$

say. Apart from the weak $T^{1/2}$ dependence, the expression for the rate coefficient k_2 is similar to the empirical *Arrhénius expression*

$$k_2 = \alpha e^{-E_\mathrm{a}/(kT)},$$

where α is a constant and E_a is an activation energy, provided that E_a is associated with the barrier energy E_1.

6.4 Photodissociation

In equation (6.4) we gave an example of photodissociation, in which the energy for a chemical reaction is provided by an incident solar photon that is absorbed by the reactant molecule (see Section 3.1). In general we can write such a reaction as

$$\mathrm{A} + h\nu \rightarrow \text{products},$$

where the number density [A] satisfies

$$\frac{\partial[\mathrm{A}]}{\partial t} = -j_{\mathrm{A}}[\mathrm{A}],$$

j_{A} being the *photodissociation rate*. This implies that, in the absence of other processes, [A] would decrease exponentially like $\exp(-j_{\mathrm{A}}t)$.

The photodissociation rate j_{A} depends upon the incident flux of solar photons, the frequency ν of the photons and the properties of the molecule A, as follows. Between frequencies ν and $\nu + d\nu$ there is a contribution

$$dj_{\mathrm{A}\nu} = \Phi_{\mathrm{A}\nu}\sigma_{\mathrm{A}\nu}\left(\frac{F_\nu^{\downarrow}}{h\nu}\right)d\nu,$$

where $\Phi_{\mathrm{A}\nu}$ is the *quantum yield* (i.e., the number of reactant molecules decomposed for each absorbed photon) and $\sigma_{\mathrm{A}\nu}$ is the absorption cross-section for the molecule A (see Section 3.5.3). $F_\nu^{\downarrow}/(h\nu)$ is the incident flux of photons (the spectral irradiance in the given frequency interval, divided by the energy per photon; see Section 3.6), assuming that the Sun is overhead and that there is no scattering, so that all incident solar photons come from above. For cases of interest to us, the quantum yield is typically quite close to a step function, being zero for frequencies less than a limit ν_0 that corresponds to the minimum energy required to dissociate the molecule and approximately unity for higher frequencies. As shown, for example, in Figure 3.15, the absorption cross-section can vary rapidly with frequency. The spectral irradiance $F_\nu^{\downarrow}$ depends on the absorbing gases present in the path of the solar beam; see Section 3.6. The total photodissociation rate is found by integration over frequency:

$$j_{\mathrm{A}} = \int_{\nu_0}^{\infty} \frac{\Phi_{\mathrm{A}\nu}\sigma_{\mathrm{A}\nu}F_\nu^{\downarrow}}{h\nu}\,d\nu.$$

There is a close relation between the processes of photodissociation and thermalization; see Section 3.1. For a given frequency interval $d\nu$, the quantum yield $\Phi_{\mathrm{A}\nu}$ equals the rate of consumption of photodissociation energy per unit volume, $dj_{\mathrm{A}\nu}\,[\mathrm{A}]h\nu$, divided by the heating rate per unit volume, $\rho Q_{\mathrm{A}\nu}\,d\nu$, say, due to absorption of photons by molecules A. This implies that photodissociation rates, like heating rates, may form Chapman layers in the vertical; see Section 3.6.

6.5 Stratospheric ozone

6.5.1 Chapman chemistry

The first attempt to explain the presence of the 'ozone layer' – that is, the region of maximum ozone number density in the lower stratosphere (see Section 1.4.1) – was made by Chapman (1930). He used a set of oxygen-only reactions, starting with the photolysis of molecular oxygen by ultra-violet photons of wavelength less than about 240 nm:

$$O_2 + h\nu \rightarrow 2O \qquad (j_2); \tag{6.11a}$$

cf. reaction (6.4). Next two fast reactions interconvert O and O_3: the termolecular reaction

$$O + O_2 + M \rightarrow O_3 + M \qquad (k_2), \tag{6.11b}$$

where M is an arbitrary air molecule (cf. reaction (6.3)); and the photolysis of ozone by photons of wavelength less than 1140 nm,

$$O_3 + h\nu \rightarrow O + O_2 \qquad (j_3). \tag{6.11c}$$

These two reactions determine the partitioning between O and O_3 within *odd oxygen*, defined as $O_x = O + O_3$. Finally, ozone is destroyed by the slow reaction

$$O + O_3 \rightarrow 2O_2 \qquad (k_3). \tag{6.11d}$$

Note that the net effect of reactions (6.11a) and (6.11b) is to produce ozone from molecular oxygen,

$$3O_2 \rightarrow 2O_3,$$

whereas the net effect of reactions (6.11c) and (6.11d) is to destroy ozone in the reverse manner,

$$2O_3 \rightarrow 3O_2.$$

The methods introduced in Sections 6.2 and 6.4 may be applied to reactions (6.11) to determine the equilibrium ozone distribution. On collecting the production and loss terms for ozone, we get

$$\frac{\partial[O_3]}{\partial t} = k_2[O][O_2][M] - j_3[O_3] - k_3[O][O_3], \tag{6.12a}$$

whereas for atomic oxygen we get

$$\frac{\partial[O]}{\partial t} = 2j_2[O_2] + j_3[O_3] - k_2[O][O_2][M] - k_3[O][O_3]. \tag{6.12b}$$

A similar equation can be written down for $\partial[O_2]/\partial t$, but it is simpler to note that, since the total number of oxygen atoms is conserved by these reactions,

$$[O] + 2[O_2] + 3[O_3] = \text{constant}. \tag{6.12c}$$

Note also that equations (6.12a) and (6.12b) can be added to give an equation for the rate of change of odd oxygen, O_x:

$$\frac{\partial[O_x]}{\partial t} = \frac{\partial[O]}{\partial t} + \frac{\partial[O_3]}{\partial t} = 2j_2[O_2] - 2k_3[O][O_3]. \tag{6.13}$$

In the steady state we set the time derivatives in equations (6.12a), (6.12b) and (6.13) to zero. These equations are not independent, but from two of them, say (6.12a) and (6.13), we can get

$$k_2[O][O_2][M] = j_3[O_3] + k_3[O][O_3], \tag{6.14a}$$

$$j_2[O_2] = k_3[O][O_3]. \tag{6.14b}$$

In the stratosphere it turns out that $j_3 \gg k_3[O]$, so equation (6.14a) gives the equilibrium ratio of the odd-oxygen species in terms of the rate coefficients j_3 and k_2 and the number densities of O_2 and air (molecule M):

$$\frac{[O]}{[O_3]} = \frac{j_3}{k_2[O_2][M]}. \tag{6.15}$$

This ratio is determined by the fast reactions (6.11b) and (6.11c); it is found to be very small in the lower and middle stratosphere, so atomic oxygen is much less abundant than is ozone there. (The opposite is true in the upper mesosphere.)

Substitution of equation (6.15) into equation (6.14b) then gives the equilibrium number density of ozone in terms of the number densities of molecular oxygen and air and the rate coefficients:

$$[O_3] = [O_2]\left(\frac{j_2 k_2[M]}{j_3 k_3}\right)^{1/2}. \tag{6.16}$$

A calculation of the vertical profile of the equilibrium ozone number density predicted by the Chapman theory can therefore be performed if the vertical profiles of the terms on the right-hand side of equation (6.16) are known. It is, however, found that the Chapman theory predicts an ozone maximum that is too large compared with observations (see Problem 6.5). Deficiencies in the Chapman theory include the neglect of further ozone-destroying reactions, particularly catalytic cycles, and the neglect of transport of ozone and other chemicals by atmospheric motions. These are discussed in Sections 6.5.2 and 6.6, respectively.

It is also important to consider the time scales on which equilibrium in the various Chapman reactions is attained. The full details are quite complicated, but we note here that, at 30 km altitude, in the mid-stratosphere, the time scale for equilibration of the odd oxygen ratio (6.15) is a few minutes, whereas the time scale for destruction of odd oxygen is several weeks. There is thus rapid adjustment between O and O_3, but only very slow changes in O + O_3.

6.5.2 Catalytic cycles

As noted in the previous section, the Chapman theory predicts too much ozone in the lower stratosphere. A major reason for this is that it neglects the effects of ozone-destroying *catalytic cycles.* These are now considered briefly.

A common catalytic cycle involves the following pair of reactions:

$$X + O_3 \rightarrow XO + O_2 \qquad (k_4), \qquad (6.17a)$$

$$XO + O \rightarrow X + O_2 \qquad (k_5). \qquad (6.17b)$$

The net effect of these two reactions is

$$O + O_3 \rightarrow 2O_2, \qquad (6.18)$$

which destroys ozone just like the Chapman reaction (6.11d). The molecule X is a catalyst; that is, it takes part in the reactions (6.17) but is not itself consumed. Instead it is 'recycled', so a single X molecule can destroy many ozone molecules and even tiny amounts of the catalyst may have important effects. Now it was noted in Section 6.5.1 that reaction (6.11d) is slow; if, however, the net effect (6.18) of the catalytic cycle is faster, this will dominate the destruction of ozone, and significantly lower the equilibrium amount (see Problem 6.6). Examples of catalysts X that actually do this in the stratosphere are the hydroxyl radical (OH), nitric oxide (NO) and chlorine (Cl).

Inclusion of catalytic cycles brings theoretical calculations of the ozone distribution into closer agreement with observations. However, the cycles interact with one another, so their effects cannot be added linearly. Further complications include the facts that some of the catalysts may become involved in additional 'null cycles', making them unavailable to reactions (6.17). Moreover, some potential catalysts may be temporarily tied up in unreactive 'reservoir species', such as dinitrogen pentoxide (N_2O_5), hydrochloric acid (HCl) and chlorine nitrate ($ClONO_2$). Comprehensive models of stratospheric chemistry must take all of these effects into account and hence may include tens of reactants and hundreds of reactions.

6.6 The transport of chemicals

It was noted above that a full theory of stratospheric ozone needs to take account of the transport of chemicals by air motion; in particular, we must apply the equations governing chemical reactions to moving parcels of air. An important concept here is the material, or advective, derivative D/Dt, introduced in Section 4.2, which represents the rate of change with time following a moving parcel.

Consider an air parcel of unit mass, including a chemical constituent A whose mass mixing ratio is μ_A. Then by definition of the mass mixing ratio (see Section 2.2), μ_A is also the mass of A in the parcel. If A is neither being produced nor lost through chemical reactions, the mass of A is constant following the parcel, so

$$\frac{D(\mu_A)}{Dt} = 0. \tag{6.19}$$

A chemical whose mixing ratio satisfies equation (6.19) is an example of a *conservative tracer*[†].

It is straightforward to show that the mass mixing ratio of A is related to the number density [A] by

$$\rho\mu_A = m_A[A], \tag{6.20}$$

where ρ is the density of air, including the constituent A (see Section 2.2), and m_A is the molecular mass of A. Equation (6.19) can therefore be written

$$\frac{D}{Dt}\left(\frac{m_A[A]}{\rho}\right) = 0$$

and so, since m_A is constant,

$$\frac{1}{\rho}\frac{D[A]}{Dt} - \frac{[A]}{\rho^2}\frac{D\rho}{Dt} = 0.$$

However, from the mass-continuity equation (4.8), $D\rho/Dt = -\rho\nabla\cdot\boldsymbol{u}$, where $\boldsymbol{u}$ is the fluid velocity. After some rearrangement and expansion of the material derivative $D/Dt = \partial/\partial t + \boldsymbol{u}\cdot\nabla$, we get

$$\frac{\partial[A]}{\partial t} + \boldsymbol{u}\cdot\nabla[A] + [A]\nabla\cdot\boldsymbol{u} = 0,$$

or, collecting the terms in $\boldsymbol{u}$,

$$\frac{\partial[A]}{\partial t} + \nabla\cdot(\boldsymbol{u}[A]) = 0. \tag{6.21}$$

This shows how the number density of a conservative tracer A changes in response to compression or expansion of a parcel of air, as it is advected with the local velocity field $\boldsymbol{u}$. The analogy with the mass-continuity equation (4.3) should be noted; see Problem 6.8.

Now suppose that A is undergoing chemical reactions so that, in addition to this compression or expansion effect, there is a contribution $-R_A$ to the rate of change of the number density [A];

[†] Other examples of conservative tracers are the potential temperature θ (when the diabatic heating $Q = 0$; see Section 4.10) and the potential vorticity P (when both the diabatic heating and friction are zero; see Section 5.1).

cf. equation (6.6). Equation (6.21) must then be replaced by

$$\frac{\partial[\mathrm{A}]}{\partial t} + \nabla \cdot (\boldsymbol{u}[\mathrm{A}]) = -R_{\mathrm{A}}. \tag{6.22}$$

All time derivatives of number densities in Sections 6.2–6.5 must be replaced by expressions analogous to that on the left-hand side of equation (6.22), when transport is to be taken into account. The right-hand side of equation (6.22) is commonly split into 'production' and 'loss' terms in the form $-R_{\mathrm{A}} = P_{\mathrm{A}} - L_{\mathrm{A}}$.

It can be shown that, when chemical reactions are taking place, equation (6.19) for the mixing ratio must be replaced by

$$\frac{D(\mu_{\mathrm{A}})}{Dt} = -\frac{m_{\mathrm{A}} R_{\mathrm{A}}}{\rho}. \tag{6.23}$$

In this case A is called a *non-conservative tracer.*

By analogy with the chemical lifetimes defined in Section 6.2, we can also define a *dynamical lifetime*

$$\tau_{\mathrm{dyn}} = \frac{[\mathrm{A}]}{|\nabla \cdot (\boldsymbol{u}[\mathrm{A}])|}.$$

This may be difficult to estimate without quite detailed knowledge of the velocity field, but, in general, if τ_{dyn} is much longer than the chemical lifetime τ_{A}, then changes in A will be mainly chemical, whereas if τ_{dyn} is much shorter than τ_{A} then A will be essentially a conservative tracer.

In principle, a full treatment of atmospheric chemistry should follow the motion of each parcel of air and account for the changes of each chemical constituent within the parcel, using equation (6.22) or equation (6.23). In practice this would be impossibly complicated, even if the motion itself were known precisely. One feasible approach is to follow the trajectories of a small but representative set of parcels and track the chemical changes for each parcel as it moves through regions of differing composition and perhaps into and out of sunlight. Such a trajectory calculation needs accurate knowledge of the wind fields and these must usually be found from models: observations on their own are seldom sufficient for an accurate determination of winds over extensive regions.

Nevertheless, some patterns of transport are well established, at least in a qualitative sense. The most important transport structure in the stratosphere is the *Brewer–Dobson circulation*: this is a meridional circulation, i.e., a circulation in the north–south and vertical directions. (The superimposed east–west motion is much more rapid, but is conceptually less important for global transport of chemicals.) It involves upward motion of air parcels from the troposphere into the stratosphere at low latitudes, poleward motion in the lower stratosphere of each hemisphere and a gentle descent back into the troposphere in middle and high latitudes; see Figure 6.4. The upward motion in low latitudes was inferred

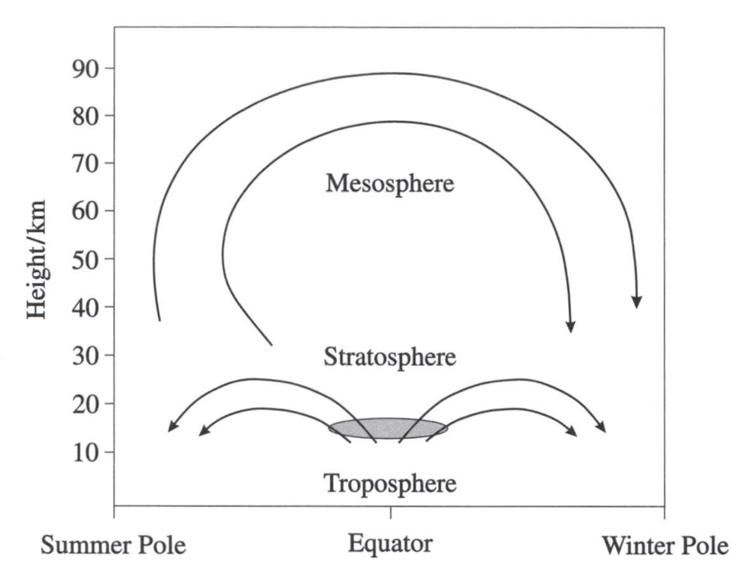

Figure 6.4 A schematic plot of the Brewer–Dobson circulation (lower four arrows, in the stratosphere) and the solstitial mesospheric circulation (upper two arrows). The shaded ellipse indicates the approximate position of the 200 K isotherm near the equatorial tropopause, i.e., the 'cold trap' identified by Brewer (1949).

by Brewer (1949) from the fact that the stratosphere is very dry. He reasoned that the inability of moisture from the troposphere to get into the stratosphere must be due to the fact that most of the air that moves from the troposphere into the stratosphere must pass through a very cold region (often called the 'cold trap' in this context) at the equatorial tropopause (see Figure 1.5), where it is 'freeze-dried', with water vapour forming ice particles that fall out under gravity. The poleward motion in the lower stratosphere was inferred by Dobson (1956) from observations of the ozone distribution: although ozone is mainly formed in the equatorial stratosphere from reactions (6.11a) and (6.11b), significant number densities are found at high latitudes, which must be due to transport from low latitudes. (Note that, if a parcel retains its ozone mixing ratio as it descends – and is compressed – at high latitudes, then the number density increases. This is also evident from equation (6.20), since the parcel encounters increasing air density as it descends.) Although the general form of the Brewer–Dobson circulation was originally obtained from the observed distributions of chemicals, elucidation and interpretation of its structure has required detailed theoretical and numerical modelling studies.

Another large-scale circulation in the middle atmosphere occurs around the solstices, with upward motion in the summer stratosphere and mesosphere, a summer-to-winter motion in the mesosphere and a descent over the winter pole; see Figure 6.4 again. This circulation also has important implications for the transport of chemicals. Both of these circulations are essentially wave-driven: the Brewer–Dobson circulation by planetary waves and the solstitial circulation mainly by gravity waves (see also Section 8.5.2).

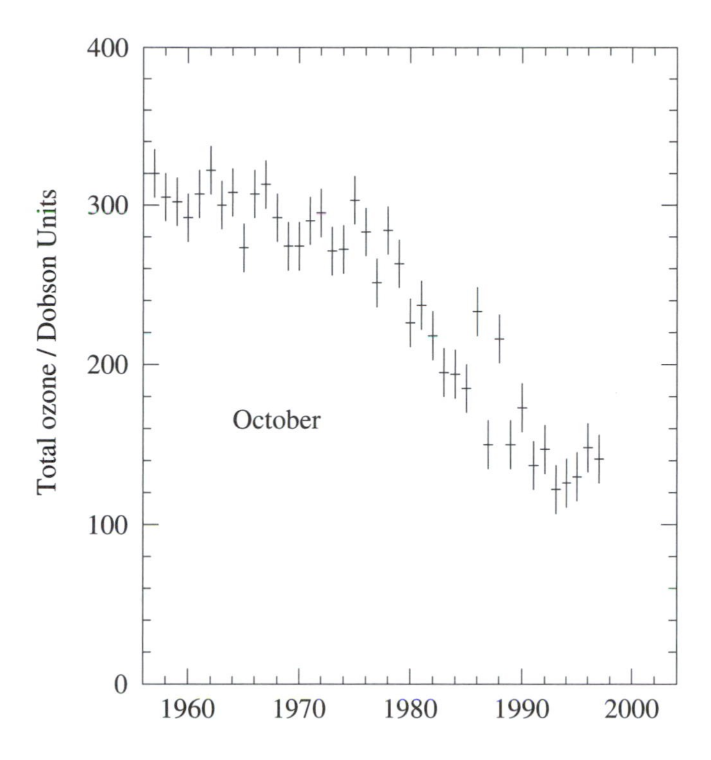

Figure 6.5 Monthly mean values of column ozone in October at Halley Bay for 1957–97. Diagram provided by Mr B. G. Gardiner.

6.7 The Antarctic ozone hole

A useful general measure of ozone in the atmosphere is the *column ozone* or *total ozone*, defined in terms of the total number N_3 of molecules of ozone in a complete vertical column of atmosphere of unit horizontal cross-section:

$$N_3 = \int_0^\infty [O_3]\, dz.$$

A convenient measure of column ozone is the *Dobson Unit* (DU), defined as the height of the column, in hundredths of a millimetre, if all the ozone molecules in it were brought to a pressure of 1 atm and a temperature of 0° C. Typical values of column ozone are about 300 DU, meaning that the height of the column, compressed in this way, would be about 3 mm! Most of the ozone molecules occur in the stratosphere.

As was mentioned in Section 1.4.4 and illustrated in Figure 1.10, the column ozone varies from place to place over the globe and with time; we here focus on its behaviour in the Antarctic. Measurements of the seasonal variation of column ozone amounts over Antarctica have been made since 1956. In particular, members of the British Antarctic Survey have made such measurements with a Dobson Spectrophotometer (see Section 7.3.1) at Halley Bay (76° S, 27° W), except during the midwinter polar night, when the Sun does

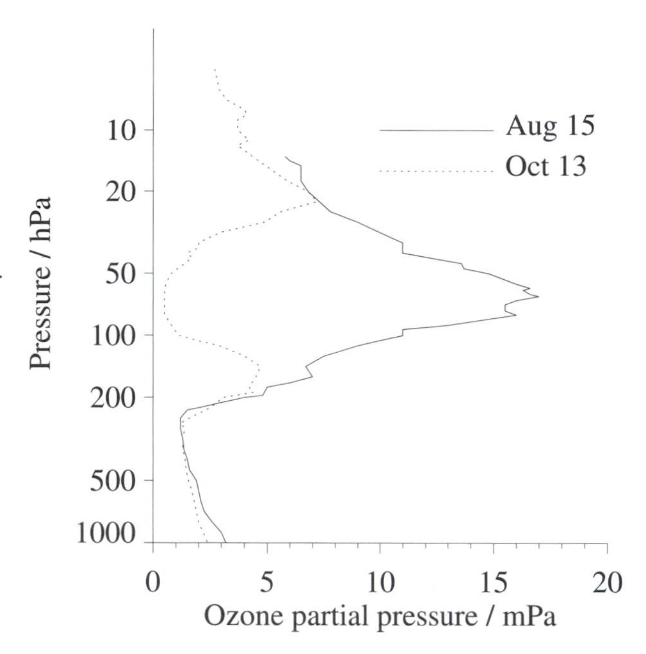

Figure 6.6 The partial pressure of ozone in the lower stratosphere over Halley Bay on 15 August (solid line) and 13 October 1987 (dotted), showing the almost total loss of ozone between about 15 and 20 km altitude during this period. Adapted after Gardiner (1989).

not rise for almost 4 months. The typical annual variation in the Antarctic, prior to the late 1970s, had a minimum of about 250–300 DU in spring and a maximum of about 400 DU in summer. However, in more recent years the picture has changed significantly, monthly mean October amounts generally decreasing throughout the 1980s and reaching values below 160 DU in the 1990s; see Figure 6.5. These routine ground-based measurements were confirmed by satellite observations with the Total Ozone Mapping Spectrometer (TOMS) and also by several special international expeditions to the Antarctic, in which very detailed measurements of the spring ozone-depletion phenomenon were made. These revealed a large-scale 'ozone hole', a region whose horizontal size approaches that of the Antarctic continent, within which massive loss of ozone occurs over a period of about 6 weeks during the Southern Hemisphere's spring, between August and October. Observations of the vertical structure of the ozone layer in the Antarctic show that, at many locations, nearly all the ozone may disappear at this time between about 15 and 20 km altitude; see Figure 6.6. Later in the year, ozone returns to the Antarctic stratosphere, although the region of depleted ozone content may drift away into lower latitudes before it is 'filled in'.

After extensive study by many atmospheric chemists and physicists a broad understanding of the Antarctic ozone hole has been attained: a qualitative description is as follows. In the winter night, a strong circumpolar (eastward) vortex forms in the Antarctic stratosphere. Unlike the Arctic polar winter vortex, this is subject to very little disturbance by planetary waves[†] and, in the absence of heat transport from lower latitude by such waves, air in the southern vortex emits infra-red radiation to space and becomes very cold. Because the vortex becomes very cold, 'polar stratospheric clouds' (PSCs) – thin clouds of ice particles – can form in the lower stratosphere.

It turns out that purely gas-phase chemistry cannot account for the large ozone losses that are found in the lower stratosphere during the Antarctic spring: 'heterogeneous' reactions, on the surfaces of the ice particles, are also necessary. (The vertical region of ozone depletion shown in Figure 6.6 is located near a PSC.) The process is too complex to be explained in detail here, but the main features are as follows.

- Catalytic cycles involving chlorine species, including the chlorine monoxide dimer Cl_2O_2, and reactive bromine species (e.g., BrO), can effectively remove ozone at the cold temperatures present in the lower Antarctic stratosphere during the spring. These cycles are inhibited if ClO becomes trapped in the reservoir species chlorine nitrate ($ClONO_2$), as noted in Section 6.5.2. However, this effect is diminished in the presence of PSCs, since nitric acid (HNO_3) condenses on the surfaces of the ice particles, which in turn may sediment out under gravity, removing the NO_2 that is needed to form $ClONO_2$.
- Moreover, the reservoir species $ClONO_2$ and HCl (hydrochloric acid) can actually release chlorine compounds (Cl_2 and HOCl) on the surface of the PSCs and these compounds are readily photolysed when the Sun returns in early spring, providing a supply of Cl for the catalytic cycles.
- A chlorine catalytic cycle of the form (6.17), where X = Cl, is too slow in the lower stratosphere to account for the observed depletion of ozone.
- Some of the stratospheric chlorine is derived from naturally occurring methyl chloride (CH_3Cl), but most comes from photolysis in the stratosphere of the man-made chlorofluorocarbons (CFCs); for example, the CFC CF_2Cl_2 is photolysed by ultra-violet radiation in the reaction

$$CF_2Cl_2 + h\nu \rightarrow Cl + CF_2Cl.$$

The Antarctic ozone hole therefore arises from a blend of chemical, dynamical and transport processes. Transport carries chlorine

[†] The Northern Hemisphere has two large continental land-masses at midlatitudes: during the winter the eastward winds blowing over these continents give rise to stationary Rossby waves in the troposphere; see Section 5.5. These in turn lead to stratospheric Rossby-wave-like disturbances that disrupt the northern winter stratospheric vortex from time to time. The absence of large midlatitude continental land-masses in the Southern Hemisphere means that such disruptions do not occur there.

compounds from the industrial regions of the Northern Hemisphere to the Antarctic. The absence of planetary waves makes the Antarctic winter stratosphere an especially cold part of the atmosphere, where PSCs can form. Finally, chemical reactions, including reactions on the ice particles of the PSCs, lead to rapid and massive local depletion of ozone in the Antarctic spring.

Ozone depletion is not confined to the Antarctic spring. Worldwide data show that there have been ozone losses at other latitudes since the 1970s: for example a 5% per decade drop in column ozone has been observed between 40° N and 60° N in winter. Possible causes of global-scale depletion include the 'dilution' of the rest of the atmosphere, due to dispersion of the Antarctic ozone hole when the vortex finally breaks down in spring; similar but weaker 'mini ozone holes' in the Arctic, which have been observed in some recent winters; and other types of heterogeneous chemistry, taking place on volcanic cloud particles and involving further chemical species.

Apart from the direct biological effects, mentioned in Section 1.4.4, the decrease in amount of atmospheric ozone may also have climatic effects. Less ozone in the stratosphere implies less absorption of solar and infra-red radiation there (see Section 3.5.3) and hence a cooler stratosphere. It is also likely that the decrease in amount of ozone during recent decades may have led to cooling in the troposphere also, partly offsetting the amplification of the greenhouse warming there due to the increase in amounts of carbon dioxide and other greenhouse gases.

References

The book by Graedel and Crutzen (1995) provides a non-technical introduction to atmospheric chemistry and its implications for climate. The basic physical chemistry mentioned in this chapter is treated in more detail by Atkins (1994) and Hobbs (1995), for example. Specialised texts on atmospheric chemistry, covering most of the topics in this chapter, include those by Wayne (2000) and Brasseur and Solomon (1986); further insights are provided by Goody (1995). Lary (1997) presents a review of ozone-destroying catalytic cycles in the stratosphere. A detailed treatment of stratospheric transport is given by Andrews *et al.* (1987). Solomon (1990) gives an authoritative account of the Antarctic Ozone Hole and some recent studies of stratospheric ozone are discussed in *Stratospheric Ozone* (1996).

Problems

6.1. Verify the result, mentioned in Section 6.1, that the photolytic reaction (6.4) requires photons of wavelength less than about

240 nm, given that the molar enthalpy of reaction is about +500 kJ mol^{-1}.

6.2. Can the reaction

$$NO_3 + H_2O \rightarrow HNO_3 + OH$$

occur spontaneously at standard temperature and pressure? Use the following Gibbs free energies of formation at STP (all in kJ mol^{-1}): HNO_3, –74.7; OH, 34.2; NO_3, 115.8; and H_2O, –228.4.

6.3. Consider a bimolecular reaction in the form of equation (6.1). Show from equation (6.7) that $[B] - [B]_0 = [A] - [A]_0$, where the subscripts 0 indicate the values at $t = 0$. Then solve for [A], assuming that the rate coefficient k_{AB} is constant. Hence verify that equation (6.8) holds if $[B]_0 \gg [A]_0$.

6.4. If $F_s = 1370\,\mathrm{W\,m^{-2}}$ is the total solar irradiance and P_s is the total solar photon flux (photons m^{-2} s^{-1}), both measured at the top of the atmosphere, show that

$$\frac{P_s}{F_s} = \frac{\int_0^\infty N(\nu)\,d\nu}{\int_0^\infty B_\nu(\nu, T)\,d\nu},$$

where $N(\nu)\,d\nu$ is the number of photons m^{-2} s^{-1} steradian^{-1} emitted from the Sun's surface in the frequency range $(\nu, \nu + d\nu)$ and $B_\nu(\nu, T)$ is the Planck function; assume that the Sun is a black body at $T = 6000$ K. Hence obtain P_s, given that

$$\int_0^\infty \frac{x^2}{e^x - 1}\,dx = 2.40.$$

Calculate also the fraction of the total photon flux that lies in each of the wavelength ranges (i) $\lambda < 1180$ nm, (ii) 240 nm $< \lambda <$ 280 nm and (iii) $\lambda <$ 240 nm. (You will need to write a computer program to calculate the required integrals.)

6.5. The following reaction rates and photodissociation rates are taken from DeMore *et al.* (1997):

$$k_2 = 6 \times 10^{-46} \left(T/300\right)^{-2.3}\ \mathrm{m^6\,s^{-1}},$$

$$k_3 = 8 \times 10^{-18} \exp\left(-2060/T\right)\ \mathrm{m^3\,s^{-1}},$$

$$j_2 = 3 \times 10^{-12}\ \mathrm{s^{-1}}, \qquad j_3 = 5.5 \times 10^{-4}\ \mathrm{s^{-1}}.$$

The photodissociation rates are appropriate for the midlatitude equinox at noon and at 25 km altitude. Use information from Figure 1.3 and Section 2.2 to estimate the number density of ozone at this altitude, based on Chapman chemistry. Show that this value is several times larger than the peak value given in Figure 1.4. What physical and chemical processes might reduce this value in the real atmosphere?

6.6. Add the catalytic reactions (6.17) to the Chapman scheme and explain how this modifies the algebra for the Chapman scheme. Assuming that the extra reactions are much slower than the fast equilibrium between O_3 and O_2+O, show that the vertical distribution of ozone is given by

$$\frac{[O_3]}{[O_2]} \approx \left(\frac{j_2 k_2[M]}{j_3 k_3}\right)^{1/2} - \left(\frac{k_2 k_4[M][X]}{2 j_3 k_3}\right);$$

compare this with equation (6.16) and comment.

6.7. What happens to O and O_3, according to the Chapman scheme, when solar radiation switches off at sunset? If catalysts are present, what happens to X and XO?

(Assume that $k_2[O_2][M] + k_3[O_3] \geq 10^{-1}\,s^{-1}$ throughout the stratosphere.)

6.8. Derive equation (6.21) by a direct method analogous to that used in Chapter 4 to derive the continuity equation (4.3).

6.9. Check that 1 DU corresponds to an ozone column of about 2.7×10^{20} molecules per square metre.

CHAPTER SEVEN

Atmospheric remote sounding

In this chapter we consider a small selection of techniques for observing the atmosphere. These techniques have been chosen for two main reasons: (a) they illustrate the use of physical principles, including principles introduced earlier in this book; and (b) they provide crucial data on atmospheric phenomena modelled elsewhere in this book, such as Rossby waves, gravity waves and the Antarctic ozone hole. The topics considered are all examples of *remote sounding*; we do not attempt to present a balanced account of all observational methods.

In Section 7.1 we briefly list some of the main atmospheric observational methods. In Section 7.2 we outline the principles of remote sounding of the atmosphere from space, focusing on methods that rely on thermal emission from atmospheric gases and on scattering of solar radiation by atmospheric gases. Then in Section 7.3 we discuss three types of ground-based remote sounding, namely the Dobson spectrophotometer, radars and lidars. We omit the details of the instruments' optical and electronic systems, the technicalities of signal processing and the sophisticated statistical methods that may be required in order to extract meaningful physical quantities from the raw measurements.

7.1 Atmospheric observations

Quantitative observations of the atmosphere are made in many different ways. Routine meteorological measurements of ground-level temperature and wind are made with simple thermometers and anemometers, respectively, and routine measurements of temperature and humidity through the depth of the troposphere are made with balloon-borne instruments (radiosondes) that transmit information back to the surface by radio. At higher altitudes, research balloons and aircraft can measure samples of air in the lower stratosphere and rockets can make measurements in the mesosphere and beyond. These are all examples of *in situ* measurements. An alternative is to use *remote sounding*, that is to measure atmospheric properties at a distance, rather than in the vicinity of the measuring apparatus. This can be done from orbiting instruments in space, from aircraft, from balloons and from ground-based instruments, in each case by measuring electromagnetic radiation emitted, scattered or transmitted by the atmosphere.

Remote-sounding techniques can be divided into *passive* and *active* types. In passive remote sounding, the radiation measured is of natural origin, for example thermal radiation emitted by the atmosphere, or solar radiation transmitted or scattered by the atmosphere. (Most space-borne remote-sounding methods are passive.) In active remote sounding, a transmitter (e.g., a radar) is used to direct pulses of radiation into the atmosphere, where they are scattered by atmospheric molecules, aerosols or inhomogeneities of atmospheric structure; some of the scattered radiation is then detected by a receiver.

Each of these techniques has its advantages and disadvantages. *In situ* measurements may give accurate, high-resolution measurements, but can sample only small, perhaps unrepresentative, regions. Remote sounding from satellites can give near-global coverage, but can provide only averaged values of the measured quantity over large regions, perhaps hundreds of kilometres in horizontal extent and several kilometres in the vertical direction. Satellite instruments are expensive to put into orbit and cannot usually be repaired if they fail. Ground-based radars can provide data with high vertical resolution (by measuring small differences in the time delays of the return pulses), but only above the radar site. In general a mix of different observational techniques must be used for providing the data required for weather forecasting or atmospheric research.

7.2 Atmospheric remote sounding from space

Earth-orbiting satellites provide enormous possibilities for global-scale measurements of the atmosphere. For example, geostationary satellites, orbiting at 36 000 km over the equator, monitor weather systems and provide information on tropospheric winds by tracking the motion of clouds. Polar-orbiting and other non-geostationary satellites, orbiting at altitudes up to about 1000 km, are used for measuring temperature and composition at a range of levels in the vertical. They complete up to 15 orbits per day, crossing each latitude circle twice per orbit; their orbits are usually chosen such that they remain 'phase-locked' with the Sun (they are then called *sun-synchronous*), crossing each latitude at the same two local times in each orbit.

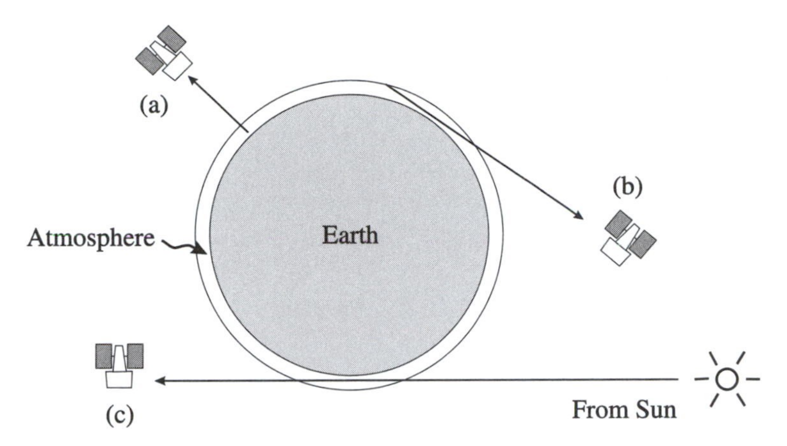

Figure 7.1 Illustrating the geometry of (a) nadir viewing, (b) limb viewing and (c) solar occultation. (Not to scale.) Note that the limb-viewing and occultation instruments may observe a different latitude from the current latitude of the satellite platform; for example, the North Pole may be viewed while the satellite is at 60° N.

In the following subsections we shall consider some basic principles of temperature and composition sounding from space, focusing on measurements of the thermal radiation emitted from the atmosphere and of solar radiation scattered by the atmosphere. Two viewing geometries are commonly used (see Figure 7.1): *nadir viewing*, in which the instrument views vertically downward towards the ground, and *limb viewing*, in which the instrument views the atmosphere tangentially, towards the limb. Nadir viewing measures radiation from a comparatively short path-length of emitting gas, against the warm background of the Earth's surface and lower atmosphere, which also emit thermal radiation. Limb viewing has the advantage of measuring radiation from a much longer path-length of emitting gas (see Problem 7.1), against the non-emitting background of cold space. However, the spacecraft must be aligned very accurately (to within a few arc seconds) for limb viewing, if the correct portion of atmosphere is to be measured. The technique of solar occultation, in which measurements of the absorption of solar radiation by the atmosphere are made as the Sun rises or sets, is also indicated in Figure 7.1, but will not be considered in this chapter. All of these techniques make use of the ideas of radiative transfer introduced in Chapter 3.

7.2.1 Thermal emission measurements

Consider a satellite instrument receiving infra-red emission from the atmosphere along a vertical (nadir) path. From the solution of the radiative transfer equation in the form of equation (3.13), assuming local thermodynamic equilibrium so that $J_\nu = B_\nu$, we can show that the spectral radiance L_ν at frequency ν received by the

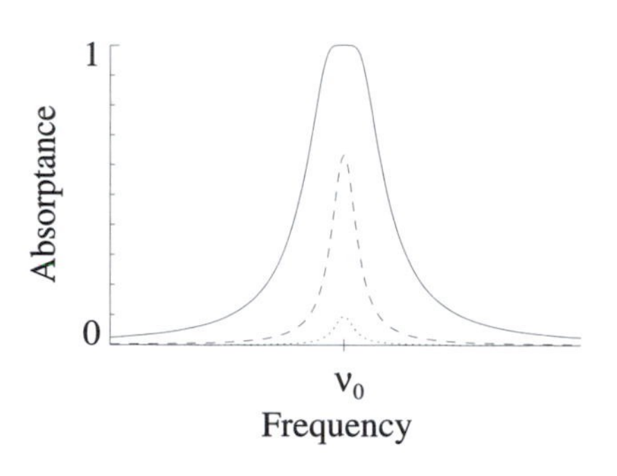

Figure 7.2 Schematic form of the spectral absorptance $\mathcal{A}_\nu(z)$ between height z and a remote-sounding satellite, as a function of frequency ν near a Lorentz line centre at ν_0, for several values of z. Solid curve, small z; dashed curve, intermediate z; dotted curve, large z.

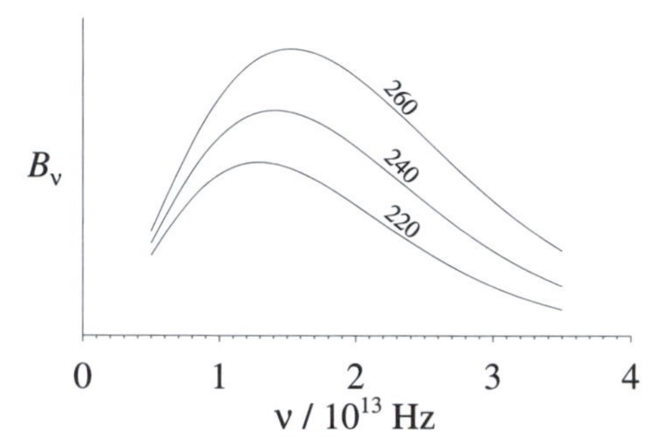

Figure 7.3 Schematic form of the Planck function $B_\nu(T)$, as a function of frequency ν, for the temperatures $T =$ 220, 240 and 260 K.

satellite is given by

$$L_\nu = \int_0^\infty B_\nu(T(z)) \frac{\partial \mathcal{T}_\nu(z,\infty)}{\partial z} dz + B_\nu(T_s)\mathcal{T}_\nu(0,\infty), \qquad (7.1)$$

where $T(z)$ is the atmospheric temperature at height z, $\mathcal{T}_\nu(z,\infty)$ is the spectral transmittance between height z and the satellite and T_s is the surface temperature (not necessarily equal to the temperature at the bottom of the atmosphere, $T(0)$). It is assumed that the surface acts as a black body†.

The spectral absorptance between height z and the satellite is $\mathcal{A}_\nu(z,\infty) = 1 - \mathcal{T}_\nu(z,\infty)$: for simplicity the '$\infty$' argument of $\mathcal{A}_\nu$ will be dropped from now on. The general form of $\mathcal{A}_\nu$ near a spectral line at $\nu = \nu_0$ is given by Figure 7.2 (cf. Figure 3.12). The precise details depend on whether we have a Doppler or Lorentz line shape, on the amount and distribution of absorbers or scatterers in the path between height z and the satellite and on the temperature and pressure distributions along the path.

The general form of the ν dependence of the Planck function $B_\nu(T)$, a much more slowly-varying function of ν than the absorptance, is as given in Figure 7.3. Note that $\partial B_\nu/\partial T > 0$ for all values of ν and T.

† Note the resemblance of equation (7.1) to equation (3.33); however the latter involves integration over all slanting upward paths.

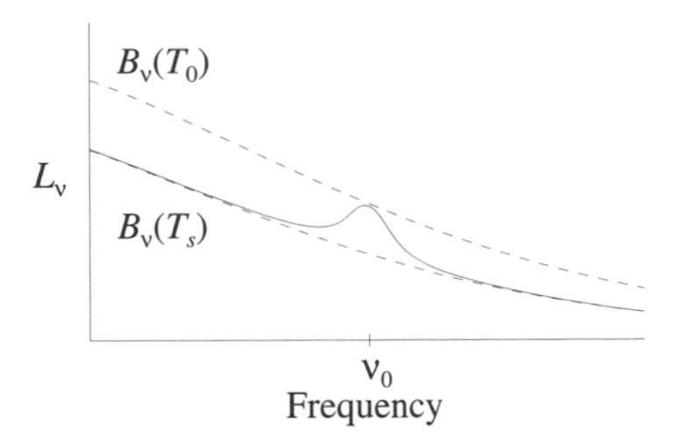

Figure 7.4 Schematic satellite-observed radiance (solid line) near a Lorentz-broadened spectral line, for an isothermal atmosphere that is warmer than the surface temperature. Planck functions at the temperatures T_s and T_0 are given as dashed lines.

The simplest case of remote sounding of temperature occurs when atmospheric extinction can be neglected. Then $\mathcal{T}_\nu = 1$ everywhere and the satellite just 'sees the surface', with no atmospheric interference, so $L_\nu = B_\nu(T_s)$. The surface temperature T_s can then be obtained from the radiance measurement L_ν by inverting the Planck function (3.1):

$$T_s = \frac{h\nu}{k \ln\left(1 + \dfrac{2h\nu^3}{c^2 L_\nu}\right)}. \tag{7.2}$$

More generally, we can define a *brightness temperature* in terms of the radiance L_ν using equation (7.2), even in the presence of extinction.

Now suppose that the frequency ν is near a spectral line centred at ν_0, so that absorption cannot be neglected, and consider an isothermal atmosphere at a constant temperature T_0, which may be different from the temperature T_s of the ground. Then the integral in equation (7.1) can be evaluated explicitly, to give

$$\begin{aligned} L_\nu &= B_\nu(T_0)[1 - \mathcal{T}_\nu(0,\infty)] + B_\nu(T_s)\mathcal{T}_\nu(0,\infty) \\ &= B_\nu(T_0)\mathcal{A}_\nu(0) + B_\nu(T_s)[1 - \mathcal{A}_\nu(0)] \\ &= B_\nu(T_s) + [B_\nu(T_0) - B_\nu(T_s)]\,\mathcal{A}_\nu(0). \end{aligned} \tag{7.3}$$

From the last line of equation (7.3) we see that the radiance L_ν equals $B_\nu(T_s)$ plus a modification equal to $B_\nu(T_0) - B_\nu(T_s)$ times the total atmospheric absorptance $\mathcal{A}_\nu(0)$. Note that any rapidly varying frequency dependence of L_ν must come from the absorptance $\mathcal{A}_\nu$, rather than from the Planck function B_ν.

Consider two cases.

- If $T_0 > T_s$ (the air temperature is greater than the surface temperature), then $B_\nu(T_0) > B_\nu(T_s)$ and L_ν displays a 'hump' around the line centre at ν_0, as shown in Figure 7.4. The line is called an *emission line.*

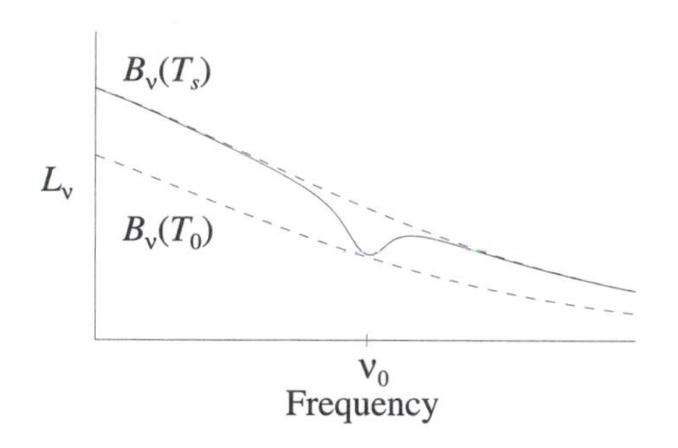

Figure 7.5 Schematic satellite-observed radiance (solid line) near a Lorentz-broadened spectral line, for an isothermal atmosphere that is colder than the surface temperature. Planck functions at the temperatures T_s and T_0 are given as dashed lines.

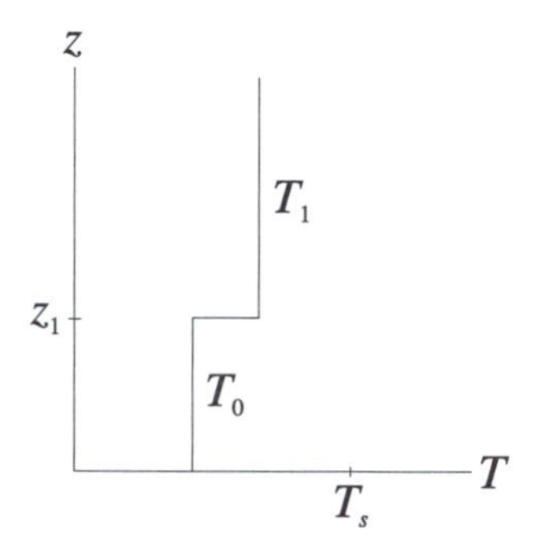

Figure 7.6 Schematic temperature profile for a two-layer atmosphere.

- If $T_0 < T_s$ (the surface is warmer than the air temperature), then L_ν displays a 'dip' around the line centre at ν_0, as shown in Figure 7.5. The line is called an *absorption line.*

In each case the radiance L_ν tends to the Planck function at the surface temperature, $B_\nu(T_s)$, sufficiently far from the line centre. The depth of the hump or dip at ν_0 in these two cases increases as the absorptance $\mathcal{A}_\nu(0)$ increases and also as the magnitude of the jump $B_\nu(T_0) - B_\nu(T_s)$ in the Planck function increases.

Now consider the two-layer atmospheric temperature profile given by

$$T(z) = T_0 \quad \text{for} \quad 0 < z < z_1, \qquad T(z) = T_1 \quad \text{for} \quad z_1 < z.$$

It is easy to show from equation (7.1) that

$$L_\nu = B_\nu(T_s) + [B_\nu(T_0) - B_\nu(T_s)]\,\mathcal{A}_\nu(0) + [B_\nu(T_1) - B_\nu(T_0)]\,\mathcal{A}_\nu(z_1).$$

Suppose for example that $T_s > T_1 > T_0$, as in Figure 7.6, crudely representing a hot surface, cool 'troposphere' and warm 'stratosphere'. The frequency variation of the resulting radiance near ν_0 received at the satellite takes the form of Figure 7.7, in the case of a Lorentz-broadened spectral line. L_ν comprises a 'background' term given by the Planck function at the surface temperature, $B_\nu(T_s)$. This has a large dip superimposed: large, because $\mathcal{A}_\nu(0)$ is a broad, high hump, and a dip because $B_\nu(T_0) < B_\nu(T_s)$. In the middle of

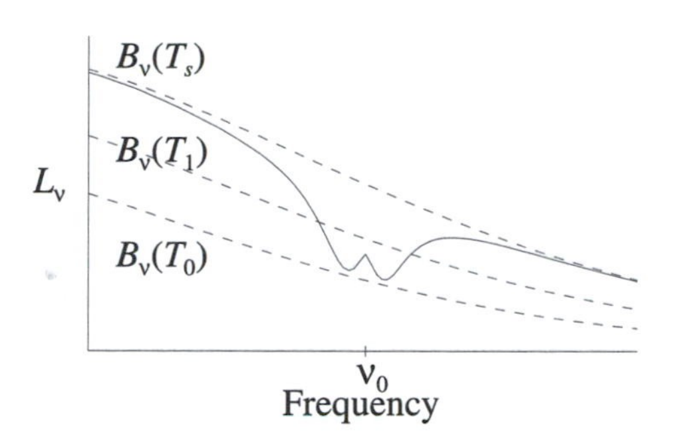

Figure 7.7 Schematic satellite-observed radiance (solid line) for the two-layer atmosphere shown in Figure 7.6. Planck functions at the temperatures T_s, T_0 and T_1 are given as dashed lines. It is assumed that $\mathcal{A}_\nu(0) \approx 1$ and $\mathcal{A}_\nu(z_1) \approx 0.6$ at the central frequency ν_0.

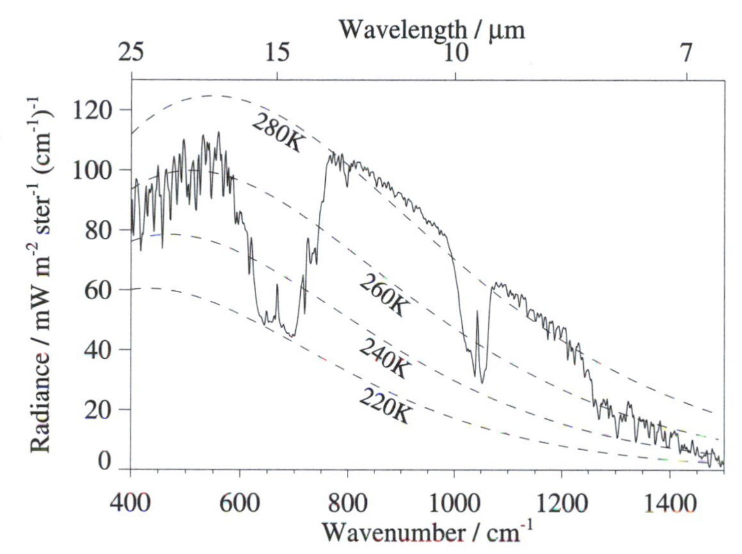

Figure 7.8 Infrared radiance measurements over the Mediterranean Sea from the IRIS instrument on the Nimbus 4 satellite. Thermal emission from the atmosphere and the surface (solid curve) is shown as a function of wavenumber (in units of cm^{-1}) and wavelength (in units of μm), together with the Planck function at several temperatures (dashed curves). Adapted after Hanel *et al.* (1971).

this large dip is a small hump: small because $\mathcal{A}_\nu(z_1)$ is generally lower and narrower than $\mathcal{A}_\nu(0)$ and also because the jump in B_ν at z_1 is less than at the ground, and a hump because $B_\nu(T_1) > B_\nu(T_0)$†. This kind of feature can be seen in the atmospheric example shown in Figure 7.8; see also Problem 7.3.

A similar approach can be used for a multi-layer atmosphere, approximating a smoothly varying profile $T(z)$ by a series of steps. If there are many layers, however, it is difficult to draw simple insights from the resulting spectral data; we therefore adopt a different approach, as follows. For convenience, we first introduce the log-pressure coordinate

$$Z = \ln(p_0/p), \tag{7.4}$$

where p_0 is the surface pressure. Recall from equation (2.15) that, for

† The wideness of the dip and narrowness of the hump result from the temperature and pressure dependences of the Lorentz width γ_L: see equation (3.22).

an *isothermal* atmosphere at temperature T_0, the pressure decreases exponentially with height, $p = p_0 \exp(-z/H)$, where $H = RT_0/g$ is the pressure scale height ($H \approx 7.6$ km for $T_0 = 260$ K). Therefore in this isothermal case $Z = z/H$. Although Z is not precisely proportional to height for a non-isothermal atmosphere, it is still approximately so in the lower and middle atmospheres. Z is sometimes called the 'log-pressure in scale heights'.

On changing variables from z to Z in the integral in equation (7.1), we obtain

$$L_\nu = \int_0^\infty B_\nu(T(Z))\,\mathcal{K}(Z)\,dZ + B_\nu(T_s)\mathcal{T}_\nu(0,\infty), \tag{7.5}$$

where

$$\mathcal{K}(Z) = \frac{\partial \mathcal{T}_\nu(Z,\infty)}{\partial Z} \tag{7.6}$$

is the *weighting function*. $\mathcal{K}$ weights the contributions to L_ν from the Planck functions at different log-pressure altitudes Z.

In certain simple cases, the weighting function can be evaluated explicitly. First note from the definition (3.23) of the transmittance that, in terms of height,

$$\mathcal{T}_\nu(z,\infty) = \exp\left(-\int_z^\infty k_\nu(z')\rho_a(z')\,dz'\right),$$

where k_ν is the extinction coefficient and ρ_a is the density of the absorbing gas. Consider the special case in which k_ν *is independent of height* and *the absorber has a constant mass mixing ratio* $\mu_a = \rho_a(z)/\rho(z)$, where ρ is the total atmospheric density. Then, using the hydrostatic relation (2.12), the transmittance is found to take the simple form

$$\mathcal{T}_\nu(z,\infty) = \exp\left(-\frac{k_\nu \mu_a p}{g}\right) \equiv e^{-\xi p},$$

say, where

$$\xi = k_\nu \mu_a / g.$$

Note that ξp is the optical depth from the satellite down to pressure p; see Section 3.6. From equations (7.4) and (7.6) the weighting function is given in terms of p by

$$\mathcal{K} = -p\,\frac{\partial \mathcal{T}_\nu(z,\infty)}{\partial p} = \xi p e^{-\xi p}, \tag{7.7}$$

or in terms of Z by

$$\mathcal{K}(Z) = \xi p_0 \exp - \left(Z + \xi p_0 e^{-Z}\right). \tag{7.8}$$

Note that the weighting function (7.8), like the heating rate in equation (3.32), has a Chapman layer structure: see Figure 7.9. From equation (7.7) it is easy to show that the weighting function takes

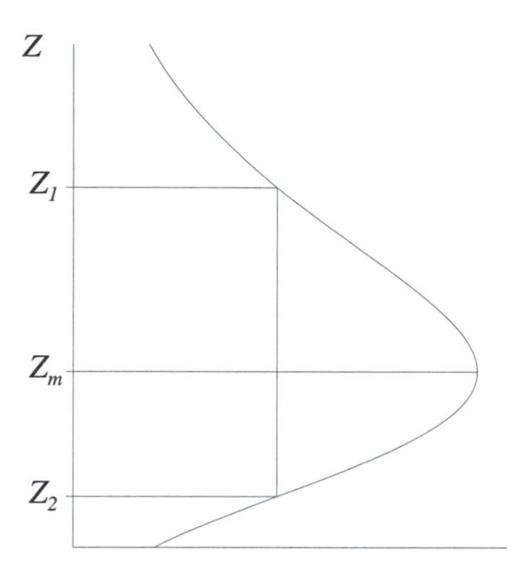

Figure 7.9 A plot of the weighting function (7.8) as a function of the log-pressure coordinate Z, for an absorber of constant mass mixing ratio and an extinction coefficient that is independent of height, for the case $\xi p_0 = 4$, i.e., $p_{\rm m} = 0.25 p_0 \approx 250$ hPa. In this case the maximum of the weighting function is at $Z_{\rm m} = 1.39$ and the points of half-maximum value are at $Z_1 = 2.85$ (corresponding to $p_1 \approx$ 58 hPa) and $Z_2 = 0.40$ (corresponding to $p_2 \approx$ 670 hPa). For other values of $p_{\rm m}$ the whole plot shifts up or down as a function of Z, without distortion.

a maximum value of e^{-1} at a pressure $p = p_{\rm m} \equiv 1/\xi = g/(k_\nu \mu_{\rm a})$ and at $Z = Z_{\rm m} = \ln(\xi p_0) = \ln(k_\nu \mu_{\rm a} p_0/g)$, corresponding to unit optical depth. The integral in equation (7.5) therefore tends to be dominated by the Planck function in the neighbourhood of this level and so gives a measure of the average brightness temperature for this region. The thickness of the region depends on how sharply peaked the weighting function is. To investigate this we consider the pressures at which $\mathcal{K}$ equals half its maximum, namely $1/(2e)$: from equation (7.7) these are given by the equation

$$\xi p e^{-\xi p} = \frac{1}{2e}.$$

With $\xi p = x$, this reduces to the equation $2x = \exp(x - 1)$, which is found by graphical or other methods to have the two roots $x =$ 0.23 and 2.68. The half-maximum points therefore occur at pressures $p_1 = 0.23 p_{\rm m}$ and $p_2 = 2.68 p_{\rm m}$, corresponding to $Z_1 = Z_{\rm m} + 1.46$ and $Z_2 = Z_{\rm m} - 0.99$, and a width in Z of $Z_1 - Z_2 = 2.45$ scale heights, a thickness of about 20 km.

Note that $p_{\rm m}$ is inversely proportional to the extinction coefficient k_ν and that $Z_{\rm m}$ is linear in $\ln(k_\nu)$. If measurements of the radiance are made at various frequencies ν, corresponding to different k_ν, different vertical regions of the atmosphere therefore dominate the integral in equation (7.5) (although the widths of the regions are still 2.45 scale heights). This fact is the basis of a standard method of remote sounding: temperature measurements of different vertical regions of the atmosphere can be inferred from radiance measurements at different frequencies. However, the broad nature of the weighting function is a major disadvantage: temperature measurements averaged over a depth of 20 km are not necessarily very useful. Narrower weighting functions can be obtained by considering cases in which the extinction coefficient is not independent of height (see Problem 7.6) or by using limb rather than nadir sounding, in which radiance measurements are made along a tangent path through the atmosphere, rather than vertically downwards; see Figure 7.1. An example of a stratospheric temperature field, measured by a limb-sounding instrument, is given in Figure 1.8.

In practice, infra-red remote sounding based on measurements from isolated parts of single spectral lines is not feasible, owing to poor signal-to-noise ratios. Therefore averages over many lines must be taken, in such a way that desirable characteristics of the weighting function are preserved.

Infra-red remote sounding of temperature in the Earth's atmosphere invariably uses emission from carbon dioxide, whose mixing ratio is nearly independent of height in the lower and middle atmosphere. Once the temperature profile $T(z)$ has been measured, a variant of the approach described above allows measurements of composition, with the absorber density ρ_a (which appears in the transmittance $\mathcal{T}_\nu$) being regarded as the unknown quantity in equation (7.1).

7.2.2 Backscatter measurements

We now consider a different type of remote-sounding technique for measuring atmospheric composition from space. The idea is to measure solar ultra-violet radiation that has been directed back to space by Rayleigh scattering[†] from atmospheric molecules. The simplest case, in which both the incoming solar beam and the scattered beam are vertical, is shown in Figure 7.10. First consider scattering from a layer of depth dz of atmosphere at height z. If the downward spectral radiance at frequency ν of solar radiation at the top of the atmosphere is $L^{\downarrow}_{\nu\infty}$, then the downward spectral radiance at height z is

$$L_\nu(z) = L^{\downarrow}_{\nu\infty} e^{-\chi_\nu(z)}, \tag{7.9}$$

where χ_ν is the optical depth,

$$\chi_\nu(z) = \int_z^\infty k_\nu(z')\rho_a(z')\,dz' \tag{7.10}$$

(see equation (3.29)), k_ν being the extinction coefficient and ρ_a the absorber density. The radiance scattered upwards from the layer is then of the form

$$L_\nu(z)\rho(z)s_\nu(z)\,dz$$

[†] See Sections 3.1 and 7.3.2.

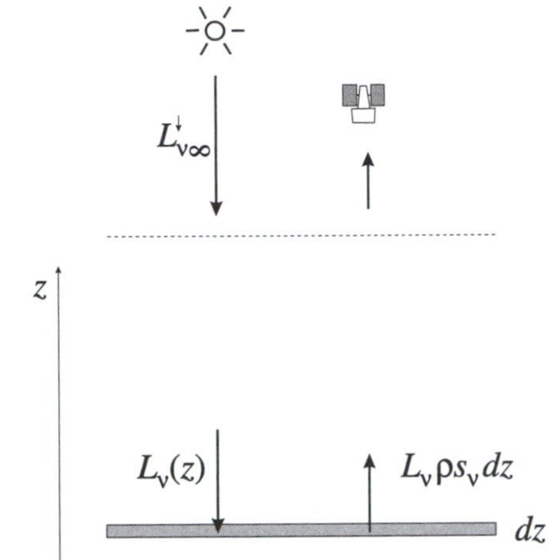

Figure 7.10 Illustrating the simplest case of backscattering of solar radiation to space, in which the incoming and backscattered beams are vertical. Only the backscattering from the shaded layer, of thickness dz, is indicated.

(cf. equation (3.8)), where ρ is the total atmospheric density and s_ν is a scattering coefficient. This radiance is attenuated by a further factor $\exp(-\chi_\nu(z))$ as the scattered beam travels back to space, so that the spectral radiance scattered from this layer that is measured at the satellite is $L^{\downarrow}_{\nu\infty}\, \rho(z)s_\nu(z)\, dz\, \exp(-2\chi_\nu(z))$. Integrating over all layers, we therefore find that the total backscattered spectral radiance measured at the satellite is

$$L^{\mathrm{s}}_\nu = L^{\downarrow}_{\nu\infty} \int_0^\infty \rho(z)s_\nu(z)e^{-2\chi_\nu(z)}\, dz. \tag{7.11}$$

Now suppose that k_ν is independent of z; then from equation (7.10) the optical depth is

$$\chi_\nu(z) = k_\nu \int_z^\infty \rho_{\mathrm{a}}(z')\, dz' = k_\nu m_{\mathrm{a}}(z), \tag{7.12}$$

where $m_{\mathrm{a}}(z)$ is the total mass of absorber above height z. If the mass mixing ratio μ_{a} of the absorber were constant, $\bar{\mu}_{\mathrm{a}}$, say, then we would have $m_{\mathrm{a}}(z) = \bar{\mu}_{\mathrm{a}} p/g$, where p is the pressure at height z. Suppose that in fact the mixing ratio is not quite constant, so that

$$m_{\mathrm{a}}(z) = \frac{\bar{\mu}_{\mathrm{a}} p}{g} + m'_{\mathrm{a}}(z), \tag{7.13}$$

where m'_{a} is small. Then from equations (7.12) and (7.13),

$$e^{-2\chi_\nu(z)} = e^{-2k_\nu \bar{\mu}_{\mathrm{a}} p/g} e^{-2k_\nu m'_{\mathrm{a}}} \approx e^{-2k_\nu \bar{\mu}_{\mathrm{a}} p/g}(1 - 2k_\nu m'_{\mathrm{a}})$$

if $k_\nu m'_{\mathrm{a}} \ll 1$. If s_ν is independent of z and the integration variable is changed from z to p, equation (7.11) becomes

$$L^{\mathrm{s}}_\nu = \frac{L^{\downarrow}_{\nu\infty} s_\nu}{g} \int_0^{p_0} [1 - 2k_\nu m'_{\mathrm{a}}(p)]e^{-\xi_1 p}\, dp,$$

where

$$\xi_1 = \frac{2k_\nu \bar{\mu}_a}{g}.$$

Then, transforming to the log-pressure variable Z defined in equation (7.4), we get

$$L^s_\nu = \frac{L^\downarrow_{\nu\infty} s_\nu}{g} \int_0^\infty [1 - 2k_\nu m'_a(Z)]\mathcal{K}_1(Z)\, dZ,$$

where

$$\mathcal{K}_1 = pe^{-\xi_1 p} = \xi_1 p_0 \exp[-(Z + \xi_1 p_0 e^{-Z})]$$

is a weighting function with similar vertical structure to that for emission measurements, given in equations (7.7) and (7.8). The measured radiance L^s_ν therefore gives an estimate of the quantity $(1 - 2k_\nu m'_a)$ near the altitude where the weighting function is a maximum. On making measurements at various frequencies, corresponding to different weighting function peaks, the vertical variation of m'_a and hence of the mixing ratio deviation $\mu_a - \bar{\mu}_a$, can be estimated. An example of an instrument that uses the backscatter technique is the Total Ozone Mapping Spectrometer, which has provided valuable information on the Antarctic ozone hole; see Section 6.7.

7.3 Atmospheric remote sounding from the ground

In this section we consider one example of passive ground-based remote sounding, namely the Dobson ozone spectrophotometer, and two examples of active ground-based remote sounding, namely radars and lidars.

7.3.1 The Dobson ozone spectrophotometer

Ground-based passive remote sounding is usually dominated by the tropospheric signal, so that stratospheric properties may be difficult to measure. A notable exception is the amount of ozone, which has its highest concentrations in the stratosphere. This fact is exploited by the Dobson ozone spectrophotometer, which measures the column ozone, the total number of ozone molecules in a column of atmosphere of unit cross-section; see Section 6.7. Consider a case when the Sun is at an angle θ from the vertical; θ is called the *zenith angle.* If horizontal variations of the absorber density in the solar beam are negligible, scattering is ignored and the extinction coefficient is independent of position, then the solar

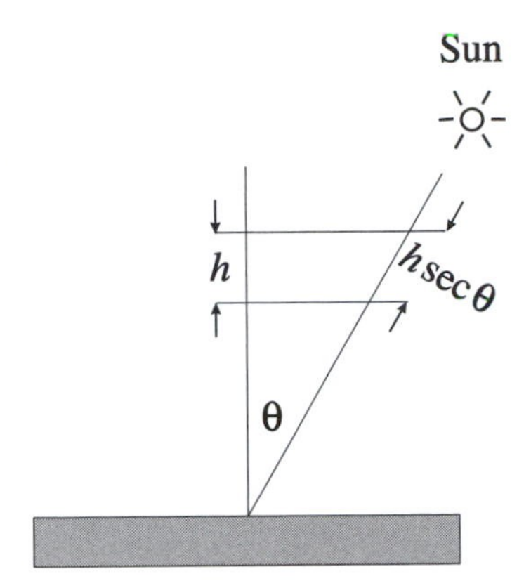

Figure 7.11 Illustrating how a non-zero solar zenith angle gives a greater path length for the solar beam. A path h for the overhead Sun becomes $h \sec\theta$ when the Sun is at zenith angle θ. The ground is shown shaded.

spectral radiance at frequency ν at the ground is

$$L_\nu = \tilde{L}_{\nu\infty} e^{-\chi_{\nu 0} \sec\theta} \tag{7.14}$$

(cf. equation (7.9)), where $\tilde{L}_{\nu\infty}$ is the solar spectral radiance along the line of sight from the Sun, at the top of the atmosphere, and

$$\chi_{\nu 0} = k_\nu \int_0^\infty \rho_\mathrm{a}(z')\, dz' \equiv k_\nu m_0 \tag{7.15}$$

is the optical path between the ground and space and m_0 is the total mass of absorber in a vertical column of unit area. The factor $\sec\theta$ in equation (7.14) allows for the increase in path length when the Sun is not overhead; see Figure 7.11.

From equations (7.14) and (7.15) we get

$$L_\nu = \tilde{L}_{\nu\infty} e^{-k_\nu m_0 \sec\theta};$$

likewise at another frequency ν' but at the same solar zenith angle,

$$L'_\nu = \tilde{L}_{\nu'\infty} e^{-k_{\nu'} m_0 \sec\theta}.$$

Therefore, after division and taking logarithms,

$$\ln\left(\frac{L_\nu}{L_{\nu'}}\right) = \ln\left(\frac{\tilde{L}_{\nu\infty}}{\tilde{L}_{\nu'\infty}}\right) + (k_{\nu'} - k_\nu) m_0 \sec\theta. \tag{7.16}$$

The technique used in measuring column ozone is to choose two frequencies in the ultra-violet, one with weak ozone absorption (small k_ν) and one with strong absorption (large $k_{\nu'}$), and carry out measurements at several zenith angles. Then, if the left-hand side of equation (7.16) is plotted against $\sec\theta$, a set of points close to a straight line is found. The slope of the line of best fit is $(k_{\nu'} - k_\nu)m_0$ and, if the extinction coefficients are known, the total mass of absorber (ozone in this case) in a vertical column is obtained. From this the column ozone is readily found.

Dobson exploited this idea in his ozone spectrophotometer, developed during the 1920s and 1930s, using simple but effective experimental techniques, including a null method for measuring $L_\nu/L_{\nu'}$. Essentially the same instrument is still used world-wide for monitoring amounts of column ozone and it was from a long series of observations with such an instrument that the Antarctic ozone hole was discovered; see Section 6.7.

7.3.2 Radars

Radars employed for atmospheric observation measure the backscatter of pulses of radio waves. Weather radars utilise backscattering of radio waves (with wavelengths of a few centimetres) from water drops and ice crystals in the troposphere to estimate precipitation rates, thus providing important data for weather forecasting systems. In the lower and middle atmosphere, backscattering of waves with a wavelength of a metre or so can take place from inhomogeneities in the atmospheric refractive index $\mathcal{N}$ for radio propagation. These inhomogeneities may, for example, be due to fluctuations of temperature and humidity associated with patches of turbulence, stratification of atmospheric properties in thin layers and regions of ionisation (including ionisation due to meteor trails). The length scales of inhomogeneities that give rise to strong scattering are of the order of half the radio wavelength or greater†. In this section we shall focus on backscattering from inhomogeneities in $\mathcal{N}$, rather than from precipitation.

Suppose that a radar transmits waves in pulses of duration τ, separated by much longer time intervals, and, for simplicity, assume that the pulses are of 'top-hat' ('box-car') shape; see Figure 7.12(a). In the case of scattering from a single target at a distance, or *range*, r_1 (see Figure 7.12(b)), the scattered pulse detected by the receiver (taken to be at the same location as the transmitter) is of the same duration and shape as the transmitted pulse, if dispersion is neglected. It is clear from Figure 7.12(b) that the time at which the beginning of the pulse is received is $t_1 = 2r_1/c$, where c is the speed of light. However, if scattering takes place at *all* ranges (see Figure 7.12(c)), then echoes from all ranges between $r_1 - \Delta r$ and r_1 reach the receiver at time t_1, echoes from all ranges between r_1 and $r_1 + \Delta r$ reach the receiver at time $t_1 + \tau$ and so on, where $2\,\Delta r = c\tau$. Therefore, at any instant between t_1 and $t_1 + \tau$ echoes are received from a layer of thickness $\Delta r = c\tau/2$, called the *range resolution*. The greatest amount of power received in this time interval originates from range r_1 and the least from $r_1 \pm \Delta r$. The range resolution

† Cf. Bragg X-ray diffraction from crystals, for which constructive interference occurs when $2d\sin\theta = n\lambda$, where d is the interplanar spacing, θ is the grazing angle, λ is the wavelength and $n \geq 1$ is an integer. From this equation it follows that $d \geq \lambda/2$. For radar scattering we have a similar formula except that d is interpreted as the length scale of the inhomogeneities in $\mathcal{N}$.

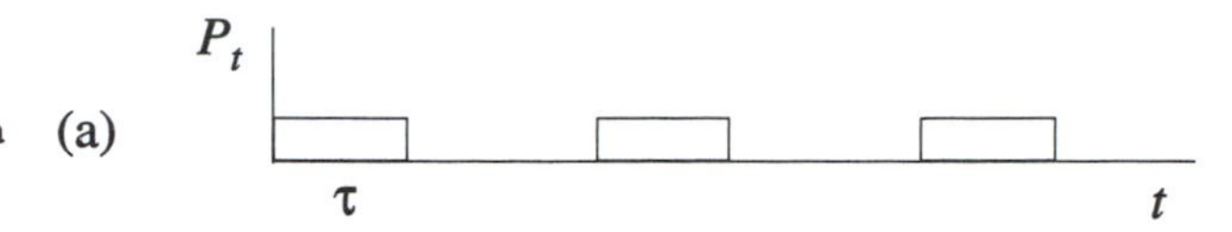

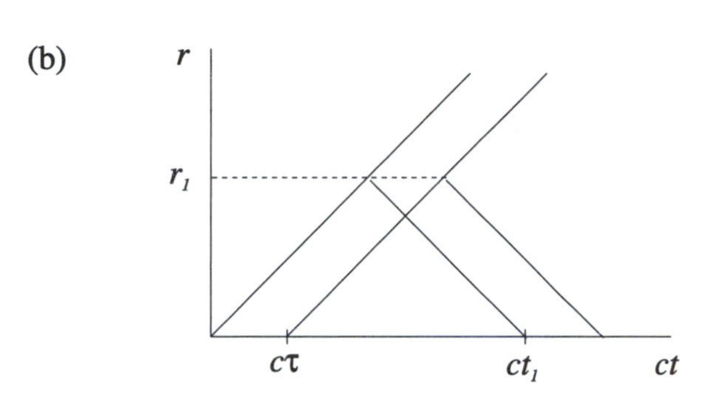

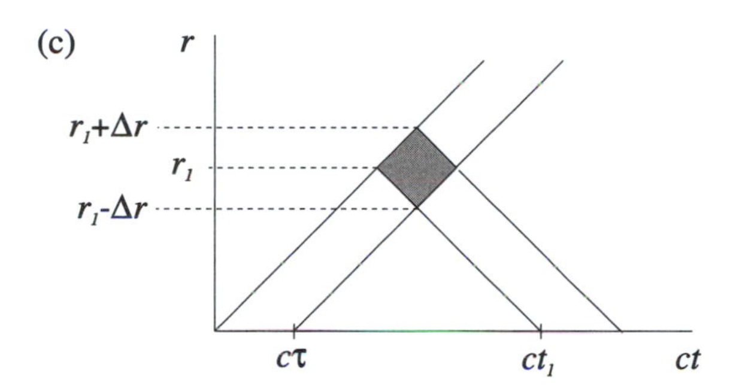

Figure 7.12 (a) A sketch of the power P_t transmitted by a radar as a function of time, showing the pulses of duration τ, assumed to be of 'top-hat' shape. (b) A plot of range r against ct, for the case of a single target at range r_1, indicating the paths of transmitted and reflected pulses. Because of the scaling of the axes, these paths have slope ± 1, which makes it clear that $t_1 = 2r_1/c$. (c) The same as (b), but for the case in which scattering takes place at all ranges. The shaded square indicates the time-varying volume of scatterers giving echoes received between times t_1 and $t_1 + \tau$. Since the diagonals of the square are equal, $2\,\Delta r = c\tau$.

gives the spatial resolution on which atmospheric quantities can be measured: for typical pulse lengths of a few microseconds it is a few hundred metres.

An important concept in scattering theory is the *differential scattering cross-section, $d\sigma/d\Omega$*. Consider scattering from a single object, as in Figure 7.13. A plane wave, carrying an energy flux (power per unit area) F say, is incident on the object and a set of scattered waves, in general propagating outwards in all directions, is produced. Then the differential scattering cross-section is defined as the scattered power $P(\boldsymbol{n})$ per unit solid angle in a direction $\boldsymbol{n}$, divided by the incident flux F:

$$\frac{d\sigma}{d\Omega} = \frac{P(\boldsymbol{n})}{F}.$$

In the case of electromagnetic waves of frequency ν scattering off an object whose dimensions are much smaller than a wavelength $\lambda = c/\nu$, the differential scattering cross-section satisfies *Rayleigh's*

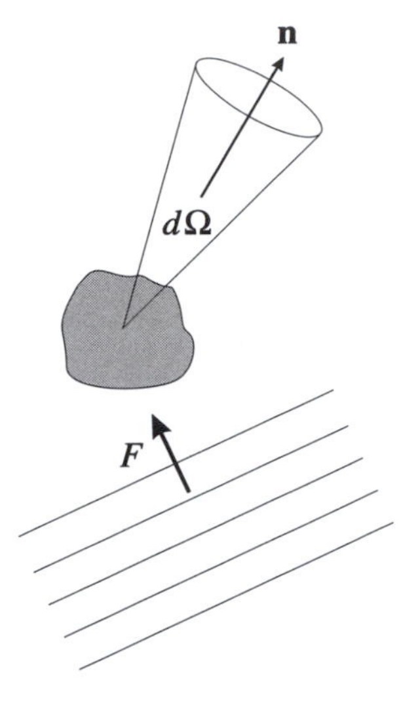

Figure 7.13 Illustrating the scattering of a plane wave, carrying flux F, by an object (shaded). Power $P(\boldsymbol{n})\,d\Omega$ is scattered into solid angle $d\Omega$ in the direction $\boldsymbol{n}$.

law,

$$\frac{d\sigma}{d\Omega} \propto \frac{1}{\lambda^4} \propto \nu^4; \tag{7.17}$$

proofs of this equation are given in advanced texts†.

Consider the simple, idealised, case of a radar transmitting power P_t in the form of spherical waves within a cone of solid angle $\alpha < 2\pi$. Suppose that a single isotropic scatterer of constant differential scattering cross-section $d\sigma/d\Omega$ is present at range r on the axis of the cone; the flux F incident on the scatterer is $P_t/(\alpha r^2)$. If the radar receiver has an antenna area A, perpendicular to the axis of the cone, it subtends a solid angle A/r^2 at the scatterer. It follows that the scattered power received by the radar is

$$P_r = \frac{P_t}{\alpha}\left(\frac{d\sigma}{d\Omega}\right)\frac{A}{r^4}. \tag{7.18}$$

(The transmittance at radio frequencies may be taken to be unity.) Note the r^{-4} dependence of P_r; this radial dependence would apply for scattering from an aircraft, for example.

Now suppose instead that the beam is filled with n scatterers per unit volume, each of differential scattering cross-section $d\sigma/d\Omega$, between ranges r and $r+\Delta r$, where $\Delta r = c\tau/2$ is the range resolution, τ being the pulse length. The beam need not be vertical; if it is

† Note that equation (7.17) implies that blue light is scattered more than red light. The same result applies to a random distribution of scatterers, which fact was applied by Rayleigh to the scattering of light from atmospheric molecules. Away from the direct solar beam, scattered blue light will predominate, leading to the blue colour of the clear sky. On the other hand, the direct beam itself will appear reddened, since much of the blue spectral component is scattered out; this is especially noticeable at sunset, when the atmospheric path-length of the solar beam is greater than that at midday.

pointed at a zenith angle θ it can be shown from equation (7.18) that the received power P_r is given in terms of the transmitted power P_t by

$$P_r = P_t \frac{A \cos\theta}{r^2} \beta \eta \, \Delta r, \tag{7.19}$$

where A is the area of the radar array, β is a volume scattering coefficient and η is an efficiency and geometric factor for the radar. If the scattering is due to turbulence, the scattering coefficient β may depend in quite a complicated way on turbulent fluctuations of the refractive index $\mathcal{N}$. If the scattering is from random thermal motion of free electrons in the ionised part of the atmosphere above about 60 km altitude, then $\beta \approx n_e \sigma_e / 2$, where n_e is the free-electron number density and σ_e is the backscattering cross-section of an electron. Note that the radial dependence in equation (7.19) is r^{-2}, in contrast to the r^{-4} dependence in equation (7.18); this is because the scattering volume is itself proportional to r^2.

An alternative situation is one in which reflection of vertically emitted radio waves takes place from an extensive horizontal layer in which the refractive index varies rapidly with height. In this case

$$P_r = P_t \frac{A^2}{r^2} \frac{\eta'}{\lambda^2} \mathcal{R},$$

where η' is another efficiency and geometric factor, $\mathcal{R}$ is the power reflection coefficient for the layer and λ is the radio wavelength. This is an example of *partial* or *specular* reflection. The dependence of $\mathcal{R}$ on the vertical gradient of the refractive index can be calculated in some cases, the simplest being that in which there is a jump $\Delta\mathcal{N}$ at a given height; see Problem 7.9.

An expression for the refractive index for radio waves of angular frequency ω is

$$\mathcal{N} - 1 = a_1 \frac{e}{T^2} + a_2 \frac{p}{T} - \frac{n_e q_e^2}{2\epsilon_0 m_e \omega^2}, \tag{7.20}$$

where p is the atmospheric pressure, e is the partial pressure of water vapour, T is the temperature, n_e is the number density of free electrons, q_e is the electronic charge, m_e is the mass of the electron, ϵ_0 is the permittivity of free space and a_1 and a_2 are constants. The first term on the right-hand side of equation (7.20) results from bound electrons present in density fluctuations of water vapour: this is most important in the lower troposphere, where water-vapour mixing ratios are highest. The second term (which is proportional to the air density, by equation (2.2)) results from bound electrons present in density fluctuations of dry air and is most important in the upper troposphere and the stratosphere. The third term results from free electrons and is most important above about 50 km, where the free electron density increases rapidly with

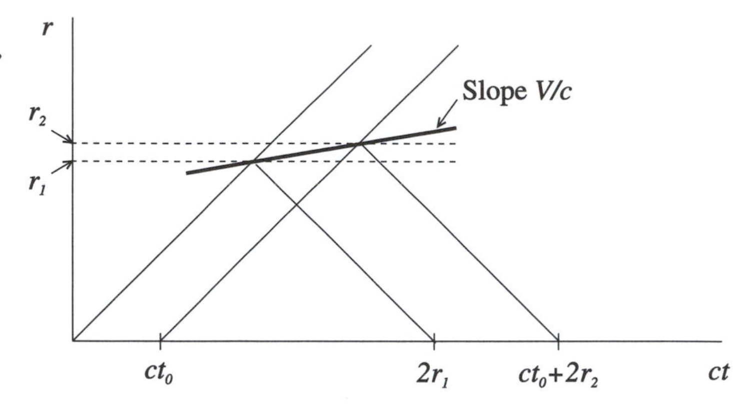

Figure 7.14 A plot of radial distance r against ct, illustrating the Doppler effect for radar measurements. The paths of two transmitted pulses and their reflections are shown. The bold sloping line shows the path of the scattering region; its slope is greatly exaggerated, since $V \ll c$. The Doppler effect is demonstrated by the fact that the time difference between the returning pulses is greater than that between the transmitted pulses. See the text for further details.

height. Each of the terms on the right-hand side of equation (7.20) is much less than 1 in magnitude.

Partial reflection techniques allow the radio refractive index $\mathcal{N}$ to be measured and hence offer the possibility of estimating the atmospheric quantities on which $\mathcal{N}$ depends, such as temperature and the free-electron density.

By measuring the Doppler shift of radar echoes from turbulence, an estimate of the radial (line-of-sight) component V of the wind velocity can be obtained, assuming that the turbulent region moves with the mean wind. Consider two short radar pulses emitted at times $t = 0$ and $t = t_0$, respectively. If the first pulse reaches the scattering region at a radial distance r_1, it is received back at the radar at time $2r_1/c$; see Figure 7.14. By the time the second pulse has reached the scattering region, this region has moved. If it is now at a range r_2, consideration of the extra distance travelled by the outgoing radar wave shows that $r_2 = r_1 + Vt_0 + V(r_2 - r_1)/c$. However, since $V \ll c$ (this calculation is of course non-relativistic) we have $r_2 = r_1 + Vt_0$ to a very good approximation. The time difference between the reception of the reflected pulses is therefore $t_0 + 2(r_2 - r_1)/c = t_0(1 + 2V/c)$.

If we apply the same argument to two adjacent peaks, say, of a sinusoidal radio wave of angular frequency $\omega_0 = 2\pi/t_0$, then the Doppler-shifted angular frequency of the reflected wave is $\omega_r = \omega_0/(1 + 2V/c) \approx \omega_0(1 - 2V/c)$. Hence the change of frequency is $\delta\omega = \omega_r - \omega_0 = -2\omega_0 V/c$ and, if $\delta\omega$ is measured, the radial component V of velocity is obtained.

If the radar is pointed in a direction given by the unit vector $\boldsymbol{n}$, then $V = \boldsymbol{n} \cdot \boldsymbol{u}$, where $\boldsymbol{u}$ is the wind vector. By varying the direction $\boldsymbol{n}$ and assuming that $\boldsymbol{u}$ does not vary significantly between the different scattering regions thus observed, the separate horizontal and vertical components of the wind can be inferred.

A variant on the above method is to measure the Doppler shift of radio waves scattered by patches of ionisation produced by meteors entering the atmosphere. These give wind estimates at altitudes between 80–100 km.

Another method for measuring the horizontal components of velocity of a moving scattering region is the spaced antenna technique. As the scattering region moves through the transmitted radar beam, it produces a moving diffraction pattern at the ground, which is sampled by a number of spaced receivers. Cross-correlation of the received signals allows the time delays between each of the receivers, and hence the required velocity, to be measured. As with the Doppler method, it is necessary to assume that the scattering region is moving with the wind.

Wind measurements of these types have been very important in determining the structure of gravity waves (see Section 5.4) in many parts of the atmosphere. An example of the use of the Doppler technique is the observation by Balsley *et al.* (1983) of an inertia–gravity wave, shown in Figure 1.7. The magnitude of turbulent fluctuations about the mean wind can be obtained from the Doppler width of the echoes.

7.3.3 Lidars

A lidar† consists of a vertically directed laser that transmits pulses of radiation, typically at visible and ultra-violet wavelengths, into the atmosphere; the backscattered radiation is measured, after a time delay, at the same site as the transmitter. The geometric and physical principles involved are similar to those for measurement using radars, described in Section 7.3.2, except that the effects of atmospheric absorption must now be considered and scattering takes place from molecules and aerosols, rather than from refractive-index variations. If a pulse of length τ and power P_t is transmitted by the laser, then the power scattered from a layer of thickness Δz at height z and detected by the receiver is

$$P_r = P_t e^{-2\chi(z)} n(z) \left(\frac{d\sigma}{d\Omega} \right) \frac{A}{z^2} \eta \, \Delta z.$$

Here $\chi(z)$ is the optical path (at the frequency ν of the laser) between the ground and height z; $n(z)$ is the number density of the scattering species and $d\sigma/d\Omega$ is its differential backscattering cross-section; and A is the telescope area and η the optical efficiency of the receiver. The height z at which scattering takes place can be inferred from the time delay $2z/c$ between transmission and reception of the pulse, where c is the speed of light. The range resolution $\Delta z = c\tau/2$ typically lies between a few tens of metres and a hundred metres.

† Light detection and ranging.

In the case in which the scattering is Rayleigh scattering from any atmospheric molecule, $d\sigma/d\Omega = a\nu^4$ by equation (7.17), where a is a known constant. The method can then be used to obtain the total number density $n(z)$ of the atmosphere and hence the density $\rho(z) = \overline{m}n(z)$ (where $\overline{m}$ is the mean molecular mass), at height z. Rayleigh scattering is important for wavelengths less than about 3 μm and allows measurements of density from heights between about 10 and 90 km. The temperature can be obtained by use of the ideal gas equation (2.2) and the hydrostatic equation (2.12): see Problem 7.10.

In the case of scattering from aerosol particles, whose sizes may approach or exceed the laser wavelength, Rayleigh scattering does not apply, and the more complex Mie scattering is the relevant process. It is difficult to use lidar measurements to quantify aerosol concentrations, but one can instead plot the lidar backscatter ratio R, defined as the ratio of the total number of backscattered photons (from molecules and aerosols) at frequency ν to the number of backscattered photons from molecules alone:

$$R = \frac{\beta_{\text{mol}} + \beta_{\text{aerosol}}}{\beta_{\text{mol}}},$$

where $\beta_i = n_i\, d\sigma_i/d\Omega$. Layers of high aerosol content, due for example to volcanic eruption, show up as sharp peaks in $R(z)$; see Figure 7.15.

Lidars with tunable dye lasers, tuned to the frequency of specific atomic or molecular transitions, can be used to measure the vertical distribution of chemical species, including various alkali atoms (deposited in the atmosphere from meteors) and ozone. For example the $^2S_{1/2}$–$^2P_{3/2}$ transition of sodium at 589.0 nm is associated with an enhanced scattering cross-section (this is an example of *resonance scattering*), which allows the lidar measurement of amounts of sodium at heights of 80–100 km.

References

A brief discussion of remote sounding from space is given by Houghton (1986), and comprehensive treatments by Houghton *et al.* (1984) and Hanel *et al.* (1992). A description of the Dobson ozone spectrometer is given by Dobson and Normand (1957). The basic physics of atmospheric radars and some typical results are presented by Balsley and Gage (1980); for a review of lidars see Thomas (1987).

Basic texts on electromagnetic waves include those by Lorrain *et al.* (1988) and Grant and Phillips (1990). The detailed physics of Rayleigh scattering is covered in the advanced texts by Jackson (1999) and Born and Wolf (1980); a descriptive account of the

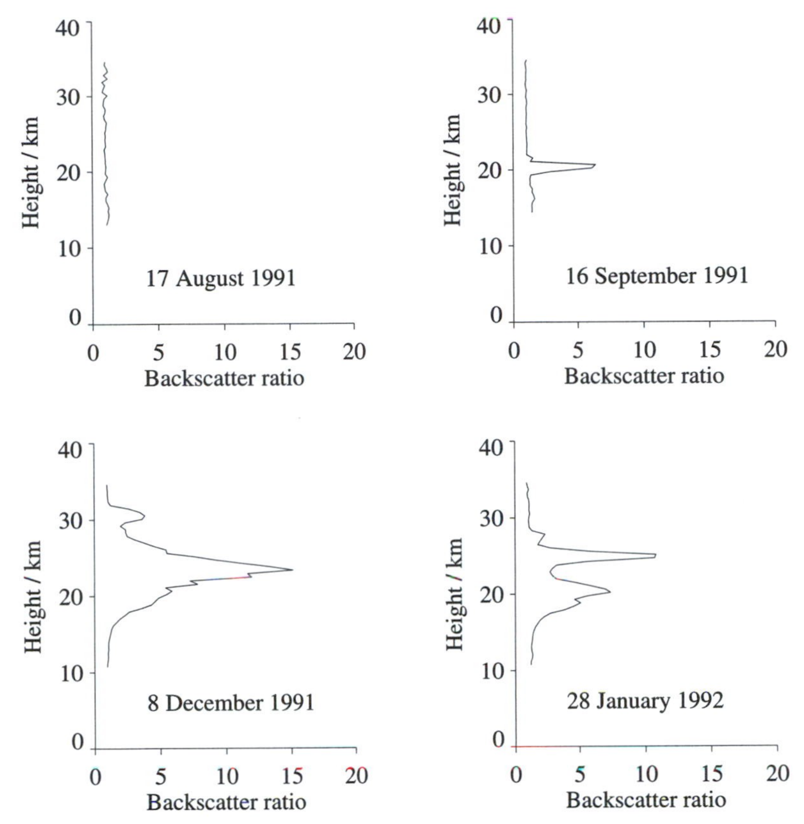

Figure 7.15 Height profiles of the lidar backscatter ratio R measured at Aberystwyth (52° N, 4° W), showing a build-up in the lower stratosphere of volcanic aerosols from Mt Pinatubo (see Vaughan *et al.* (1994)). On 17 August 1991, $R \approx 1$ throughout the profile, indicating that little aerosol is present. On 16 September, R exhibits a peak in a thin layer at about 20 km altitude. On 8 December there is a very large peak, extending over a deep layer centred at about 23 km. On 28 January 1992 a double-peaked structure is present. Based on data provided by Dr G. Vaughan.

phenomenon is given by Lynch and Livingston (1995), who also detail many other examples of atmospheric optical phenomena. Mie scattering is treated by Liou (1980).

Problems

7.1. Consider a thin layer of gas of uniform density, between altitudes z and $z + d$ in a spherical atmosphere, emitting thermal radiation. Show that the ratio of the maximum mass of the gas in a limb-viewing path to the mass of gas in a nadir-viewing path is approximately $(8a/d)^{1/2}$, where a is the Earth's radius. Evaluate this ratio for $d = 10$ km and comment on your result.

7.2. Derive equation (7.1) from equation (3.13), assuming local thermodynamic equilibrium, given that the Earth's surface is a black body.

7.3. Figure 7.16 shows emission spectra from Mars.

(*a*) What do the fine features between 50 and 25 μm, especially in the south polar spectra, indicate?

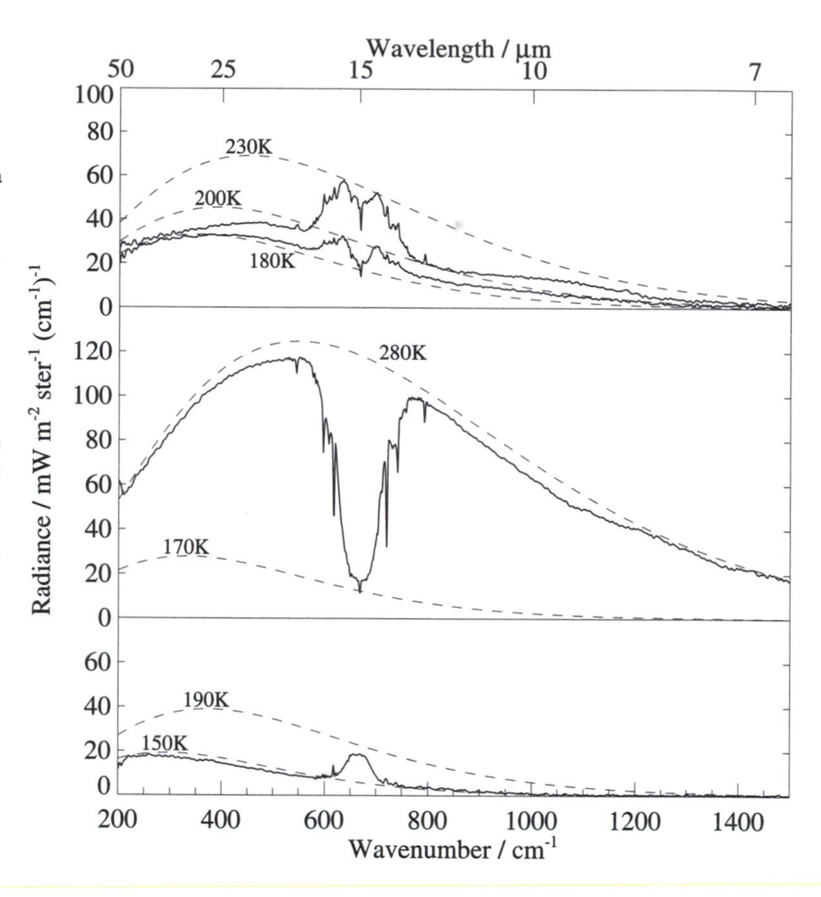

Figure 7.16 Emission spectra from Mars obtained with the IRIS instrument on Mariner 9 (adapted after Hanel *et al.* (1992)). Top panel: spectra recorded over the south polar region; upper curve includes a smaller fraction of the polar ice cap than the lower curve. Middle panel: spectrum recorded near 21° S. Lower panel: spectrum recorded near 66° N; note that the condensation temperature of CO_2 at Martian surface pressures is about 145 K. Diagram prepared with the help of Dr S. R. Lewis, using data from the Planetary Data System.

(*b*) What is responsible for the large feature centred on 15 μm?

(*c*) What can be said about the surface and atmospheric temperatures?

(*d*) What can be learnt from a comparison of terrestrial (e.g., Figure 7.8) and Martian midlatitude spectra?

7.4. Measurements of outgoing radiation are made in three spectral intervals by a nadir-viewing radiometer in Earth orbit, with the following results:

(*a*) brightness temperature 310 K (atmospheric window at 11 μm),

(*b*) brightness temperature 220 K (CO_2 band at 15 μm), and

(*c*) brightness temperature 280 K (O_3 band at 9.6 μm).

Assuming that the surface is black, that the CO_2 absorption is so strong in interval (b) that the stratosphere is black in that interval and that all the ozone resides in the stratosphere, calculate the transmission of the ozone layer at 9.6 μm. What could we learn about the ozone layer from this?

(The Planck function B_ν at 9.6 μm is approximately 5, 21 and 36 in units of 10^{-13} W m^{-2} steradian^{-1} Hz^{-1} at 220, 280 and 310 K, respectively.)

7.5. The 11 μm region is often referred to as a 'window' in the terrestrial atmosphere, but there is still some absorption, mainly due to water vapour. Assuming that the absorption coefficient k due to the water is $k = \alpha e$, where e is the vapour pressure and α is a constant, and that the mass mixing ratio μ at pressure level p is $\mu_0(p/p_0)^3$, where μ_0 and p_0 refer to the bottom boundary and μ_0 is small, show that the transmittance of a vertical path from level p to a satellite is $\exp[-\beta(p/p_0)^8]$ where

$$\beta = \frac{\alpha}{g}\frac{M}{M_w}\frac{\mu_0^2 p_0^2}{8}$$

and M_w and M are the molar masses of water vapour and air, respectively.

Given that $\alpha = 7 \times 10^{-6}$ m^2 kg^{-1} Pa^{-1} and that the temperature T_0 and relative humidity $\mathit{\Lambda}_0$ at the bottom boundary are 290 K and 70%, respectively, show that $\beta = 0.102$ and calculate the transmittance from the surface to the satellite. (The SVP of water vapour at 290 K is 1936.7 Pa.)

Write down an expression for the net deficit in the spectral radiance received from the surface by a nadir-viewing radiometer due to attenuation and emission by water vapour. Evaluate this deficit for the case of a (black) sea surface at temperature T_0 for the conditions given above, assuming that $T = T_0(p/p_0)^\delta$ and $B_\nu = \eta(T/300)^\gamma$. Take $p_0 = 10^5$ Pa, $\delta = 0.29$, $\eta = 4 \times 10^{-12}$ W m^{-2} steradian^{-1} Hz^{-1} and $\gamma = 4.5$.

$$\left(\int_0^1 y^{8.305} \exp(-0.102 y^8)\, dy = 0.102\right)$$

What would be the error in inferring the temperature of the sea surface from radiance measurements if the effect of the water vapour were not taken into account? Is this error significant?

7.6. What is meant by a *weighting function* in atmospheric remote sounding? Show that, if the transmittance $\mathcal{T}(p) = \exp(-\beta p^\alpha)$, where α and β are constants, and if the height-like variable $Z = \ln(p_0/p)$ is used as a vertical coordinate, the vertical weighting function is

$$\mathcal{K}(p) = \alpha\beta p^\alpha \exp(-\beta p^\alpha).$$

Find α and β for an isothermal atmosphere containing an absorber with a small and constant mass mixing ratio μ when

(*a*) the absorber is grey with constant absorption coefficient k;

(*b*) the radiometer is sensitive to a single frequency in the far wings of a Lorentz line; and (*c*) the radiometer is sensitive to a spectral interval in which the transmittance from level p to the satellite is given by the strong limit of the Goody random model,

$$\mathcal{T} = \exp\left(-\frac{2}{\delta}(Sm\gamma_{\mathrm{L}})^{1/2}\right),$$

where S is the mean line strength, m is the mass of absorber per unit area in the path, γ_{L} is the mean Lorentz half-width and δ is the mean line spacing.

7.7. Find the pressure levels at which the peaks of the weighting functions occur and find also the widths (in terms of pressure) at half maximum of the weighting functions of the previous question. Sketch the functions for cases (a) and (b).

(The roots of $2x = e^{x-1}$ are approximately 2.68 and 0.23.)

7.8. Verify the last term on the right-hand side of equation (7.20) for the radio refractive index, by ignoring the effects of bound electrons and proceeding as follows.

It can be shown (see Lorrain *et al.* (1988) or Jackson (1999)) that the dispersion relation for transverse plasma waves is

$$\omega^2 = \omega_{\mathrm{p}}^2 + c^2k^2,$$

where the *plasma frequency* ω_{p} is given by

$$\omega_{\mathrm{p}}^2 = \frac{n_{\mathrm{e}}q_{\mathrm{e}}^2}{\epsilon_0 m_{\mathrm{e}}},$$

in the notation of Section 7.3.2. Hence show that, if $\omega^2 \gg \omega_{\mathrm{p}}^2$,

$$\mathcal{N} \equiv \frac{ck}{\omega} \approx 1 - \frac{n_{\mathrm{e}}q_{\mathrm{e}}^2}{2\epsilon_0 m_{\mathrm{e}}\omega^2}.$$

7.9. Suppose that the atmosphere has a radio refractive index $\mathcal{N}(z)$ given by

$$\mathcal{N} = \mathcal{N}_1 \quad \text{for} \quad z < z_0, \qquad \mathcal{N} = \mathcal{N}_2 \quad \text{for} \quad z > z_0.$$

Consider a vertically propagating electromagnetic wave incident on the level $z = z_0$ from below. By using standard methods for calculating the reflected wave (see, e.g., Grant and Phillips (1990) or Lorrain *et al.* (1988)) show that the power reflection coefficient $\mathcal{R}$ (the ratio of the reflected to the incident power) is given by

$$\mathcal{R} = \tfrac{1}{4}(\mathcal{N}_1 - \mathcal{N}_2)^2.$$

(You may find it convenient to take $z_0 = 0$ for this calculation.)

7.10. Given observations of density $\rho(z)$ from lidar measurements and the temperature $T(z_{\mathrm{m}})$ at a single height z_{m}, use the ideal gas equation and hydrostatic balance to show that the full temperature profile $T(z)$ can be obtained from

$$T(z) = \frac{1}{\rho(z)}\left(\rho(z_{\mathrm{m}})T(z_{\mathrm{m}}) + \frac{g}{R}\int_{z}^{z_{\mathrm{m}}} \rho(z')\,dz'\right).$$

Show that the term $\rho(z_{\mathrm{m}})T(z_{\mathrm{m}}) \to 0$ as $z_{\mathrm{m}} \to \infty$. (In practice z_{m} is taken as the maximum height of the lidar measurements and a value of $T(z_{\mathrm{m}})$ is guessed; the resulting values of $T(z)$ are not very sensitive to this guess.)

CHAPTER EIGHT

Atmospheric modelling

This chapter is a short introduction to the use of models in atmospheric research and forecasting. In Section 8.1 we explain how a hierarchy of models – simple, intermediate and complex – can be used for gaining understanding of atmospheric behaviour and interpreting atmospheric data. In Section 8.2 we give brief details of the numerical methods used in the more complex theoretical models, while in Section 8.3 we outline the use of these models for forecasting and other purposes. In Section 8.4 we describe an example of a class of laboratory models of the atmosphere. Finally, in Section 8.5, we give some examples of atmospheric phenomena that arise from interactions between basic physical processes and that can be elucidated only with the aid of models of intermediate complexity.

8.1 The hierarchy of models

The basic philosophy of atmospheric modelling was outlined in Section 1.2. It was mentioned there that a hierarchy of models, from simple to complex, must be used for understanding and predicting atmospheric behaviour; this hierarchy is illustrated in Figure 8.1. The simple models ('back-of-the-envelope' or 'toy' models) involve a minimum number of physical components and are described by straightforward mathematical equations that can usually be solved analytically. These models provide basic physical intuition: most of the models considered earlier in this book are of this type. The intermediate models involve a small number of physical components but usually require a computer for solution of the mathematical equations. Simple and intermediate models can provide valuable conceptual pictures of atmospheric processes, but do not usually give accurate simulations of actual atmospheric behaviour. The complex models (often called *general circulation models* or GCMs)

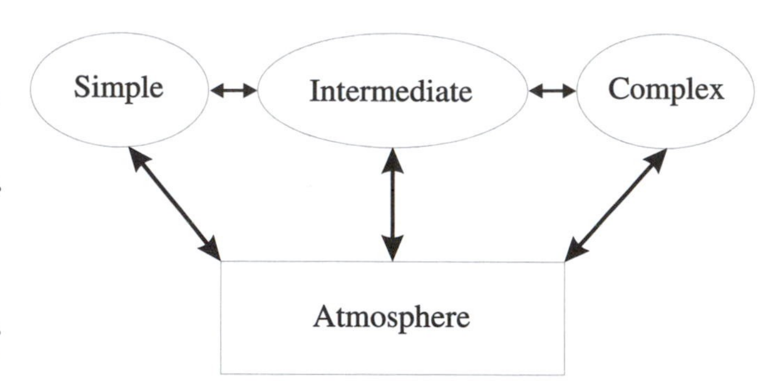

Figure 8.1 Illustrating the use of a hierarchy of atmospheric models. Each type of model is motivated by, and provides some insight into, atmospheric behaviour. The simpler models help us interpret the more complex models. If a more complex model fails, a simpler model may help us to locate the cause.

contain mathematical representations of a large number of physical processes. These are the models that are used when accurate simulations are required.

Each type of model is motivated by atmospheric observations and each may help us understand some aspect of atmospheric behaviour. The physical intuition provided by the simplest models helps us interpret the intermediate models and the intermediate models help us interpret the complex models. When a complex model fails to produce a reasonable simulation of observed atmospheric behaviour, it may be necessary to use an intermediate model (perhaps a 'stripped-down' version of the complex model) to find the cause.

The types of physical process that can be included in atmospheric models fall into three main categories, each of which has been studied in earlier chapters.

- *Dynamical processes*, including those parts of thermodynamics associated with the First Law. These were introduced in Chapters 4 and 5.
- *Radiation*: this was introduced in Chapter 3.
- *Chemistry*: some aspects of this were introduced in Chapter 6.

These three categories interact in complex ways in the atmosphere, as is partly illustrated in Figure 8.2. Transport by dynamical processes carries chemicals from one part of the atmosphere to another, for example vertically from ground level to the stratosphere, or horizontally from the industrialised Northern Hemisphere to Antarctica; see Section 6.6. The heating due to absorption of solar ultra-violet radiation by ozone may drive dynamical processes. Solar radiation may be energetic enough to disrupt chemical bonds, leading to photochemical reactions. In general models incorporating even a few of these interactions will be too complicated for simple analysis and so will be of 'intermediate' or 'complex' type.

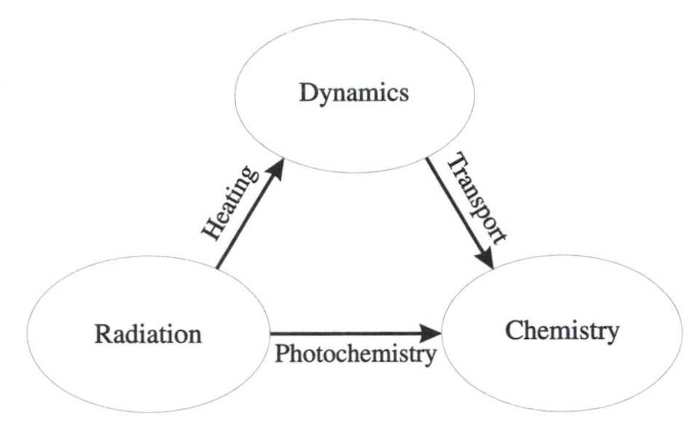

Figure 8.2 Illustrating some of the interactions among dynamics, radiation and chemistry in the atmosphere; see the text for details.

8.2 Numerical methods

As mentioned in Section 8.1, intermediate and complex models require us to solve the mathematical equations describing the physics by means of computers. This is not an easy task and may require extremely careful and efficient computer programming and huge computer codes. In this section we describe some of the basic principles involved.

The equations describing the dynamics are generally partial differential equations, involving differentiation in space and time. Consider for example the x-momentum equation; under the approximations discussed in Section 4.7.2 this can be written

$$\frac{\partial u}{\partial t} = -u\frac{\partial u}{\partial x} - v\frac{\partial u}{\partial y} - w\frac{\partial u}{\partial z} + fv - \frac{1}{\rho}\frac{\partial p}{\partial x} + F^{(x)}, \qquad (8.1)$$

by equations (4.23a) and (4.22). The physical meanings of the symbols are given in Section 4.7.2, but are not important for the present discussion. The partial time derivative of the eastward velocity u on the left-hand side of equation (8.1) can be approximated by a *finite difference*: from the Taylor expansion

$$u(x, t+\delta t) = u(x, t) + \delta t\,\frac{\partial u(x,t)}{\partial t} + O(\delta t^2), \qquad (8.2)$$

where δt is assumed small, we have

$$\frac{\partial u(x,t)}{\partial t} \approx \frac{u(x, t+\delta t) - u(x,t)}{\delta t}.$$

Now suppose that all the quantities on the right-hand side of equation (8.1) are known at time t; then this equation gives $\partial u/\partial t$ at time t and substitution into equation (8.2) gives u at time $t + \delta t$, with an error of order δt^2. A 'time-step' (in effect a forecast for a tiny time interval) has therefore been performed. In principle, repeated use of this process allows longer forecasts to be made.

The spatial derivatives on the right-hand side of equation (8.1) must also be calculated; one way to do this is again by finite differences. The atmosphere is taken to be spanned by a three-dimensional grid or lattice of points, with spacings δx, δy and δz, say, in the eastward, northward and vertical directions, respectively†. Then, for example, a simple 'centred difference' approximation to the pressure gradient $\partial p/\partial x$ is

$$\frac{\partial p(x,t)}{\partial x} \approx \frac{p(x+\delta x,t)-p(x-\delta x,t)}{2\,\delta x},$$

and other spatial derivatives can be represented in a similar way, thus allowing computation of the terms on the right-hand side of equation (8.1). For current finite-difference NWP models, the horizontal grid spacing is typically about 1° longitude by 1° latitude, with 15–30 vertical levels spaced at 50–100 hPa in pressure (and less than 50 hPa near the ground and the tropopause) and a total of about 10^6 grid-boxes.

An alternative is to calculate horizontal derivatives by *spectral methods*, that is, the expansion of physical quantities in series of orthogonal functions whose derivatives are known explicitly. Consider for example an x derivative again and focus for simplicity on an f-plane (see Section 4.7.3) that is periodic in the x direction, with a period L that represents the length of the latitude circle. A suitable expansion of $u(x,t)$ is then a Fourier series of the form

$$u(x,t) = \bar{u}(t) + \sum_{n=1}^{N}\left[a_n(t)\cos\left(\frac{2n\pi x}{L}\right) + b_n(t)\sin\left(\frac{2n\pi x}{L}\right)\right],$$

where N is called the *truncation limit*; the x derivative of u is

$$\frac{\partial u(x,t)}{\partial x} = \sum_{n=1}^{N}\left[\frac{2n\pi b_n(t)}{L}\cos\left(\frac{2n\pi x}{L}\right) - \frac{2n\pi a_n(t)}{L}\sin\left(\frac{2n\pi x}{L}\right)\right].$$

In spherical coordinates the expansion must be in terms of spherical harmonics $\exp(im\lambda)\,P_n^m(\cos\phi)$, where λ is longitude, ϕ is latitude, P_n^m is an associated Legendre polynomial and $|m| \le n$. Modern spectral NWP models include harmonics with values of n up to about 200; finite differences are still usually used in the vertical. Since the equations of motion such as equation (8.1) are nonlinear, it is necessary to re-expand products of pairs of terms, each represented by its own series, into single series. Fast numerical algorithms have been developed for this purpose.

An ever-present danger in numerical modelling is the possibility of numerical instabilities, that is, instabilities of the finite-differenced equations that are not present in the real physical system being modelled. If they are left unchecked, these instabilities may completely

† In spherical geometry special precautions must be taken at the poles. Note also that in complex models pressure coordinates (see Section 4.9) or other vertical coordinates are usually used instead of the height z.

swamp the numerical solution; special care must be taken to avoid or inhibit them.

Numerical problems in the modelling of radiation are of a different type. Here the most important quantity to be calculated is the radiative heating rate, which is known in principle from the solution to the radiative transfer equation (3.10). However, the calculation must be done for several absorbing gases, over a wide variety of wavelengths or frequencies, and for atmospheric paths at different zenith angles. This would involve extensive computations if it were carried out in full, so some of these calculations are often simplified. For example, rather than attempting to integrate in frequency over every known spectral line (the 'line-by-line' approach) one may work with a small number of band transmittances (see equation (3.25)), averaged over spectral bands containing many lines. The presence of clouds, which scatter and absorb the radiation, is a further complication.

If chemistry is to be included in a model, then coupled ordinary differential equations representing the chemical reactions (see Sections 6.2 and 6.4) will be needed. Although these appear less complicated than the partial differential equations of dynamics, there may still be difficulties in solving them numerically. For example, there may be a large number of chemicals involved and hence a large number of equations in the set. Less obviously, the possibility of both fast and slow reactions implies a disparity of natural time scales in the differential equations, which are then said to be 'stiff' and require careful numerical treatment.

8.3 Uses of complex numerical models

Over the last few decades the need for reliable weather forecasts has provided a strong impetus for the development of complex atmospheric models, at least for the troposphere and lower stratosphere. Numerical weather prediction (NWP) models, for predicting a few days to perhaps two weeks ahead, must include representations of large-scale dynamical processes, small-scale frictional processes in the atmospheric boundary layer, small-scale drag due to gravity waves, short-wave and long-wave radiative transfer, water-vapour transport, the effects of clouds, precipitation and the transfer of heat, momentum and moisture between the surface and the atmosphere. Some of these (such as gravity-wave-induced drag and also latent heating and precipitation due to cumulus clouds) take place on length scales below the grid size and are therefore not explicitly represented in the models. Such processes must be *parameterized*, that is, expressed in terms of quantities that *are* explicitly represented. Such parameterizations often involve empirical formulae that do not have rigorous physical justification and may involve disposable constants whose values are poorly known. Much of the art of NWP consists in selecting these constants in such a way as to optimise the resulting forecasts.

Forecasts are made by integrating the equations of motion forwards in time, starting from an observed initial state. Owing to measurement error and the inevitable sparsity of data in some regions, the initial state may not be known with great accuracy. This can be a major source of error in forecasts. In common with other chaotic processes, the evolution of the atmosphere can be sensitive to initial conditions. One way to estimate the reliability of the forecast model on a given occasion is to perform several forecasts, starting from slightly different initial conditions. If all members of the 'ensemble' of forecasts are similar, then a high degree of reliability can be placed on a representative forecast; however, if the ensemble contains widely differing forecasts, less weight can be placed on any of them.

Complex models are also used for climate forecasting, for years or even decades into the future. Here the long-term behaviour of the atmosphere (perhaps averaged over seasons) is of interest, rather than the detailed day-to-day evolution. On these longer time scales, physical processes that are not important for NWP must be included: for example, some representation of oceanic heat transfer will be needed. If questions like ozone depletion or the effect of increasing greenhouse gases are to be considered, chemical transport and chemical reactions must be included. Models of this kind involve very heavy usage of computer resources, not only because they must be run for many simulated years but also because of the variety of physical processes included and hence the complexity of the mathematical equations to be solved. Moreover, great care must be used in examining and interpreting the huge quantity of data produced by these models.

Complex models of the NWP type are being used increasingly for *data assimilation*. This is a procedure by which observed data (which may be incomplete, or inconsistent because of measurement error) are fed into the model over a period of time. The model can be regarded as an accurate representation of the known laws of physics and dynamics; it adjusts the observations, within their expected error bounds, in such a way as to force them to be closely consistent with these laws. The resulting 'assimilated' data set thus represents a dynamically and physically consistent interpolation of the observed evolution with time of the atmosphere over the observation period. Data assimilation is also used for deriving accurate initial conditions for weather forecasting.

Complex models are heavily used for research purposes, to help understand atmospheric behaviour. In this mode they fit particularly well into the hierarchy described in Section 8.1. A set of controlled experiments may be designed to investigate a particular

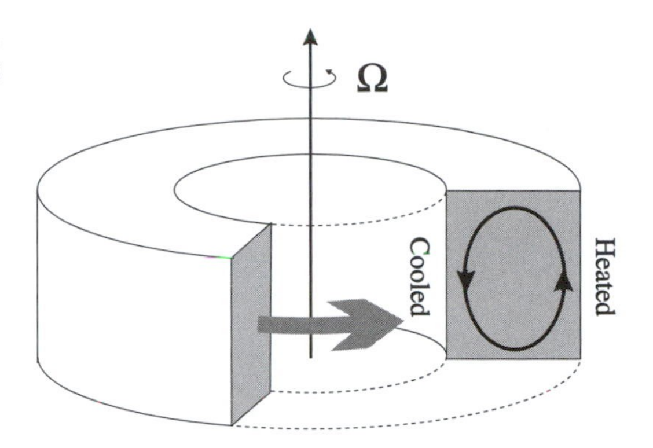

Figure 8.3 A schematic illustration of the rotating annulus, heated at the outer wall and cooled at the inner wall. A 'cut-away' view is shown, with an indication of the azimuthal, vertical and radial motion found at slow rates of rotation.

phenomenon – for example the development of the Antarctic ozone hole in a given year. Simulations of the phenomenon are performed, using various initial conditions, representations of the chemical reactions and so on. Comparison of the resulting simulations with observed data may help to identify which aspects of the initial conditions are most important for the phenomenon or which chemical schemes are closer to reality. Intermediate and simple models, for interpretation of the successful simulations and for investigation of the reasons why others are unsuccessful, may also play crucial roles in such an investigation.

A further use for complex models is in the design of large measurement programmes, by providing representative data sets on which observation strategies and data-analysis schemes can be refined, in advance of the collection of the real data.

8.4 Laboratory models

A quite different approach from those described above is to use a laboratory apparatus to model atmospheric phenomena. This has the advantage of providing a real physical analogue, free from numerical errors; it may also be subjected to detailed and systematic measurement in a way that may not be possible with the atmosphere. On the other hand, great care must be taken to minimise the effects of physical properties of the laboratory system that are irrelevant to the atmosphere.

A good example of a laboratory system that has been used extensively for modelling atmospheric processes is the *rotating annulus*, which was developed and used by R. Hide and co-workers over many years. The annulus consists of a ring-shaped container of fluid, such as water, which is for example heated at the outer wall by maintaining it at temperature T_o and cooled at the inner wall by maintaining it at temperature $T_i < T_o$. The annulus is placed on a rotating turntable and rotated at a constant angular speed

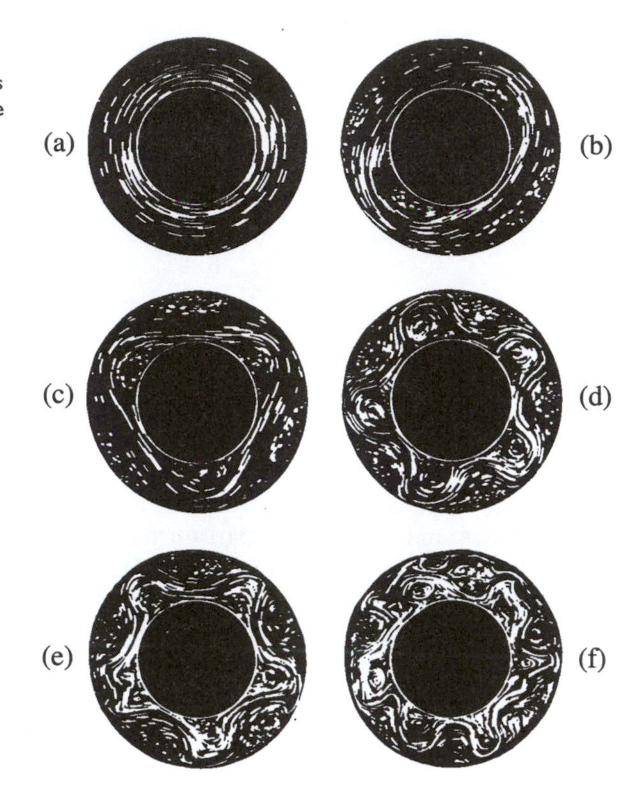

Figure 8.4 Flow regimes in a rotating annulus in which velocities near the upper surface are made visible by 'streak photography', i.e., by illuminating suspended particles with a flat light beam and taking a time exposure with a camera mounted on the rotating apparatus. The rotation rate Ω increases from frame (a) to frame (f). Note the azimuthally symmetric flow in (a), the increasing number of wavelengths in the azimuthal direction in (b) to (e) and the breakdown to irregular flow in (f). From Hide and Mason (1975).

Ω about its axis; see Figure 8.3. The fluid in the annulus mimics the atmosphere, the rotation mimics the rotation of the Earth, the heating at the outer radius mimics the heating of the atmosphere in the tropics and the cooling at the inner radius mimics the cooling at high latitudes.

The fluid flow in the annulus takes a number of forms, depending on the speed of rotation and the thermal contrast between the inner and outer walls. At slow rates of rotation, the flow is basically in the azimuthal direction (corresponding to eastward flow in the atmosphere), with a superimposed vertical and radial motion: rising on the heated outer wall, inward flow along the top, descent on the inner wall and outward flow along the base (roughly analogous to the observed mean meridional 'Hadley Circulation' in the low-latitude atmosphere). This flow has no azimuthal variation.

At faster rates of rotation the flow is no longer azimuthally symmetric, but rather forms a number of wave-like patterns that drift in azimuth (see Figure 8.4); the number of wavelengths around the annulus depends on the speed of rotation. These waves are related to the baroclinic instabilities discussed in Section 5.7. At certain rates of rotation the number of wavelengths or the amplitude of the waves may pulsate slowly: this pulsation is called *vacillation*.

At high rates of rotation, the wave-like flow breaks down to a disordered, turbulent state reminiscent of some disturbed atmospheric flows. It turns out that an analogue of the β-plane (see Section 4.7.3) can be constructed by sloping the top and bottom walls of the annulus, which allows Rossby waves to be simulated. Extensive observational and theoretical analysis has been carried out on this type of laboratory system and similar theoretical analysis has been carried out on numerical models of the annulus flow. Despite there being some obvious differences from the atmosphere (for example the much greater role of viscosity and the presence of side-walls in the laboratory analogue) the rotating annulus has provided much insight into atmospheric (and also oceanic) behaviour.

8.5 Final remarks

As noted in Section 8.1, most of the models considered in this book have been of the simplest type; we have generally concentrated on single atmospheric processes in isolation, ignoring interactions with other processes. This approach is essential for developing a basic physical intuition for the atmosphere. However, the number of atmospheric phenomena that may be described fully in terms of one type of process alone is limited, so for most purposes interactions between processes must be invoked if a reasonably complete physical description is to be provided. This usually means that models of at least intermediate complexity must be used. In this concluding section we give some examples of such physical descriptions that have been developed with the help of intermediate models.

8.5.1 The height of the tropopause

A climatological zonal-mean temperature field for January was given in Figure 1.5 and some physical processes that help determine this field were outlined in Section 1.4.1. It was noted there that even a basic understanding of the processes determining the position of the tropopause – the interface between the troposphere and stratosphere – involves the interaction between radiation and dynamics. Except in the winter polar regions and in the tropics, radiation is the primary process determining the temperature structure in the lower stratosphere; here ozone absorbs infra-red and solar radiation, causing heating which is mostly balanced by long-wave cooling. On the other hand, dynamical processes such as convection and baroclinic instability (see Section 5.7) are important for heat transport in the troposphere. In the presence of moisture, convection tends to relax the tropospheric lapse rate towards the saturated adiabatic lapse rate (see Section 2.8); the effects of baroclinic instability are more subtle. All of these processes must be taken into account if the position of the tropopause is to be modelled properly; however, although many comprehensive general circulation models produce good simulations of the tropopause height, a complete physical understanding is still lacking.

8.5.2 The middle-atmosphere temperature field

Other aspects of the zonal-mean temperature field also depend on the interplay between dynamics and radiation. It was mentioned in Section 6.6 that there is a large-scale meridional mass circulation in the middle atmosphere, sketched in Figure 6.4, that is driven by wave motions. The full details are beyond the scope of this book; suffice it to say that nonlinear and dissipative processes associated with upward-propagating gravity waves drive the summer-to-winter solstitial circulation in the upper mesosphere, including rising motion in the mesosphere over the summer pole and descent over the winter pole, and that nonlinear and dissipative processes associated with upward-propagating midlatitude Rossby waves drive the Brewer–Dobson circulation in the stratosphere. As well as being important for tracer transport, as noted in Section 6.6, these circulations also influence the temperature field. It was mentioned in Section 3.6.4 that radiative processes act rather like a 'spring', which tries to pull the temperature field towards a purely radiative equilibrium. Such an equilibrium would include warm temperatures in the summer upper mesosphere and near the tropical tropopause and cool temperatures in the stratospheric polar night. The dynamically driven meridional circulation acts against the radiative spring, forcing temperatures below equilibrium values in the regions of rising motion in the summer upper mesosphere and near the tropical tropopause, and forcing temperatures above equilibrium values in the region of descent in the winter (especially the northern winter) stratosphere.

8.5.3 The Antarctic ozone hole

The Antarctic ozone hole, described in Section 6.7, provides an example of an atmospheric phenomenon in which chemical, dynamical (particularly transport) and radiative effects are all significant. Dynamically driven transport is of course important for carrying ozone from the ozone-production regions at low latitudes to the polar regions; it also carries chlorine compounds from industrial regions to the Antarctic stratosphere. On the other hand, the polar winter vortex in the Southern Hemisphere is much less disturbed by Rossby wave activity than is the northern winter vortex, so temperatures are closer to the cold radiative-equilibrium values. At such cold temperatures polar stratospheric clouds can form, upon which the crucial heterogeneous chemical reactions that provide the ozone-destroying chlorine compounds take place. Later in the Antarctic spring, when the polar vortex breaks down (due to dynamical influences), transport may carry ozone-depleted air to lower latitudes, thus perhaps leading to a decrease in the amount of ozone over a larger region than the Antarctic alone.

References

A good introduction to numerical modelling of the atmosphere is given by Holton (1992); the book by Haltiner and Williams (1980) provides a more detailed account. The book edited by Trenberth (1992) contains a number of articles dealing with various aspects of climate modelling. Read (1993) gives a critical account of applications of the ideas of chaos theory to the atmosphere and to laboratory analogues of the atmosphere.

Recent work on the processes controlling the height of the midlatitude tropopause is reported by Thuburn and Craig (1997). The processes controlling the temperature of the cold equatorial tropopause are discussed by Holton *et al.* (1995). For the summer mesopause see Andrews *et al.* (1987).

APPENDIX A

Useful physical constants

In this appendix, STP denotes the standard temperature 273.15 K = 0 °C and pressure 1013.25 hPa = 1 atm. The hectopascal (hPa = 10^2 Pa) is used as the unit of pressure; it is equivalent to the millibar used in older works. Note that, in SI units, the gramme mole (mol) should strictly speaking be replaced by the kilogramme mole, or kilomole (kmol = 10^3 mol).

Constant	*Symbol*	*Numerical value*
Universal constants		
Universal gas constant	R^*	$8.31\ \mathrm{J\,K^{-1}\,mol^{-1}}$
or (SI value)		$8.31 \times 10^3\ \mathrm{J\,K^{-1}\,kmol^{-1}}$
Avogadro's number	N_A	$6.02 \times 10^{23}\ \mathrm{mol^{-1}}$
or (SI value)		$6.02 \times 10^{26}\ \mathrm{kmol^{-1}}$
Planck constant	h	$6.63 \times 10^{-34}\ \mathrm{J\,s}$
Boltzmann constant	k	$1.38 \times 10^{-23}\ \mathrm{J\,K^{-1}}$
Speed of light	c	$3.00 \times 10^8\ \mathrm{m\,s^{-1}}$
Stefan–Boltzmann constant	σ	$5.67 \times 10^{-8}\ \mathrm{W\,m^{-2}\,K^{-4}}$
(Note that $\sigma = 2\pi^5 k^4/(15h^3c^2)$)		
The Earth		
Mean acceleration due to gravity at the Earth's surface	g	$9.81\ \mathrm{m\,s^{-2}}$
The Earth's mean radius	a	6371 km
The Earth's mean rate of rotation	Ω	$7.29 \times 10^{-5}\ \mathrm{s^{-1}}$
Standard surface pressure	p_0	1013.25 hPa

The Sun		
Solar constant	F_s	$1370\,\mathrm{W\,m^{-2}}$
Mean distance between the Earth and the Sun		$1.50 \times 10^{11}\,\mathrm{m}$
Mean radius of the Sun		$6.96 \times 10^{8}\,\mathrm{m}$
Dry air		
Molar mass of dry air	M	$28.97\,\mathrm{kg\,kmol^{-1}}$
Density of dry air at STP	ρ_0	$1.29\,\mathrm{kg\,m^{-3}}$
Specific heat capacity of dry air at STP:		
at constant pressure	c_p	$1005\,\mathrm{J\,K^{-1}\,kg^{-1}}$
at constant volume	c_v	$718\,\mathrm{J\,K^{-1}\,kg^{-1}}$
Specific gas constant for dry air	R	$287\,\mathrm{J\,K^{-1}\,kg^{-1}}$
Water		
Molar mass of water	M_w	$18.02\,\mathrm{kg\,kmol^{-1}}$
Density of liquid water at STP	ρ_w	$1000\,\mathrm{kg\,m^{-3}}$
Density of ice at STP	ρ_i	$917\,\mathrm{kg\,m^{-3}}$
Specific heat capacity of water vapour at 0 °C:		
at constant pressure		$1850\,\mathrm{J\,K^{-1}\,kg^{-1}}$
at constant volume		$1390\,\mathrm{J\,K^{-1}\,kg^{-1}}$
Specific heat capacity of liquid water at 0 °C		$4217\,\mathrm{J\,K^{-1}\,kg^{-1}}$
Specific heat capacity of ice at 0 °C		$2106\,\mathrm{J\,K^{-1}\,kg^{-1}}$
Specific gas constant for water vapour	R_w	$461\,\mathrm{J\,K^{-1}\,kg^{-1}}$
Specific latent heat of vaporization at 0 °C	L_v	$2.50 \times 10^{6}\,\mathrm{J\,kg^{-1}}$
Specific latent heat of vaporization at 100 °C	L	$2.26 \times 10^{6}\,\mathrm{J\,kg^{-1}}$
Specific latent heat of fusion at 0 °C	L_f	$0.33 \times 10^{6}\,\mathrm{J\,kg^{-1}}$
Specific latent heat of sublimation at 0 °C	L_s	$2.83 \times 10^{6}\,\mathrm{J\,kg^{-1}}$
(Note that $L_s = L_v + L_f$.)		

Sources include Kaye and Laby (1986) and Lide (1995).

APPENDIX B

Derivation of the equations of motion in spherical coordinates

In this appendix we derive the atmospheric equations of motion (4.21) in spherical coordinates, starting with the Navier–Stokes equation (4.20). In terms of the unit vectors $\boldsymbol{i}$, $\boldsymbol{j}$ and $\boldsymbol{k}$, the velocity vector can be written as $\boldsymbol{u} = u\boldsymbol{i} + v\boldsymbol{j} + w\boldsymbol{k}$ and the rotation vector $\boldsymbol{\Omega} = \Omega(\boldsymbol{j}\cos\phi + \boldsymbol{k}\sin\phi)$. Note that the unit vector $\hat{\boldsymbol{\Omega}} = \boldsymbol{\Omega}/\Omega = \boldsymbol{j}\cos\phi + \boldsymbol{k}\sin\phi$ is fixed in space, if the small variations in the Earth's rotation are neglected.

Now $\boldsymbol{i}$, $\boldsymbol{j}$ and $\boldsymbol{k}$ change with time, following the motion, and this must be taken into account when calculating the components of $D\boldsymbol{u}/Dt$; we get

$$\frac{D\boldsymbol{u}}{Dt} = \frac{Du}{Dt}\boldsymbol{i} + \frac{Dv}{Dt}\boldsymbol{j} + \frac{Dw}{Dt}\boldsymbol{k} + u\frac{D\boldsymbol{i}}{Dt} + v\frac{D\boldsymbol{j}}{Dt} + w\frac{D\boldsymbol{k}}{Dt}.$$

The material derivatives of the unit vectors can be calculated as follows. Note first that $\boldsymbol{k} = \boldsymbol{r}/r$, where $\boldsymbol{r}$ is the position vector and r is its magnitude. Then

$$\frac{D\boldsymbol{k}}{Dt} = \frac{D}{Dt}\left(\frac{\boldsymbol{r}}{r}\right) = \frac{1}{r}\frac{D\boldsymbol{r}}{Dt} - \frac{\boldsymbol{r}}{r^2}\frac{Dr}{Dt}.$$

However $D\boldsymbol{r}/Dt = \boldsymbol{u}$ and $Dr/Dt = w$, so

$$\frac{D\boldsymbol{k}}{Dt} = \frac{\boldsymbol{u}}{r} - \frac{\boldsymbol{k}w}{r} = \frac{1}{r}(u\boldsymbol{i} + v\boldsymbol{j}). \tag{B.1}$$

Next use the fact that the unit vector $\hat{\boldsymbol{\Omega}} = \boldsymbol{j}\cos\phi + \boldsymbol{k}\sin\phi$ is constant. This implies that

$$0 = \frac{D\hat{\boldsymbol{\Omega}}}{Dt} = \cos\phi\frac{D\boldsymbol{j}}{Dt} - \frac{v\sin\phi}{r}\boldsymbol{j} + \sin\phi\frac{D\boldsymbol{k}}{Dt} + \frac{v\cos\phi}{r}\boldsymbol{k}, \tag{B.2}$$

since $D\phi/Dt = v/r$. Substitution of equation (B.1) into equation (B.2) then leads to

$$\frac{D\boldsymbol{j}}{Dt} = -\frac{u\tan\phi}{r}\boldsymbol{i} - \frac{v}{r}\boldsymbol{k}. \tag{B.3}$$

Now

$$\boldsymbol{i} = \boldsymbol{j} \times \boldsymbol{k}, \tag{B.4}$$

by orthogonality of the unit vectors. So

$$\frac{D\boldsymbol{i}}{Dt} = \frac{D\boldsymbol{j}}{Dt} \times \boldsymbol{k} + \boldsymbol{j} \times \frac{D\boldsymbol{k}}{Dt}. \tag{B.5}$$

Substitution from equations (B.1) and (B.3) into equation (B.5) and use of equation (B.4) and the analogous identity $\boldsymbol{j} = \boldsymbol{k} \times \boldsymbol{i}$, together with the identity $\boldsymbol{k} \times \boldsymbol{k} = 0$, gives

$$\frac{D\boldsymbol{i}}{Dt} = \frac{u \tan \phi}{r} \boldsymbol{j} - \frac{u}{r} \boldsymbol{k}. \tag{B.6}$$

Alternative, geometrical, derivations of equations (B.1), (B.3) and (B.6) are given in Section 2.3 of Holton (1992).

The Coriolis term in equation (4.20) can be written

$$2\boldsymbol{\Omega} \times \boldsymbol{u} = 2\Omega(w \cos\phi - v \sin\phi)\boldsymbol{i} + 2\Omega u \sin\phi\, \boldsymbol{j} - 2\Omega u \cos\phi\, \boldsymbol{k}.$$

In terms of the small incremental distances $dx = r \cos\phi\, d\lambda$ in the eastward direction, $dy = r\, d\phi$ in the northward direction and $dz = dr$ in the vertical direction, the pressure gradient term in equation (4.20) can be written

$$\nabla p = \frac{\partial p}{\partial x}\boldsymbol{i} + \frac{\partial p}{\partial y}\boldsymbol{j} + \frac{\partial p}{\partial z}\boldsymbol{k}.$$

Neglecting the centripetal acceleration, equation (4.20) can therefore be written

$$\begin{aligned}
&\frac{Du}{Dt}\boldsymbol{i} + \frac{Dv}{Dt}\boldsymbol{j} + \frac{Dw}{Dt}\boldsymbol{k} \\
&\quad + \frac{u}{r}(u \tan\phi\, \boldsymbol{j} - u\boldsymbol{k}) - \frac{v}{r}(u \tan\phi\, \boldsymbol{i} + v\boldsymbol{k}) + \frac{w}{r}(u\boldsymbol{i} + v\boldsymbol{j}) \\
&\quad + 2\Omega(w \cos\phi - v \sin\phi)\boldsymbol{i} + 2\Omega u \sin\phi\, \boldsymbol{j} - 2\Omega u \cos\phi\, \boldsymbol{k} \\
&\qquad\qquad + \frac{1}{\rho}\left(\frac{\partial p}{\partial x}\boldsymbol{i} + \frac{\partial p}{\partial y}\boldsymbol{j} + \frac{\partial p}{\partial z}\boldsymbol{k}\right) + g\boldsymbol{k} = \boldsymbol{F},
\end{aligned}$$

where $\boldsymbol{F} = F^{(x)}\boldsymbol{i} + F^{(y)}\boldsymbol{j} + F^{(z)}\boldsymbol{k}$ is the frictional force $\eta\nabla^2\boldsymbol{u}/\rho$. Collecting the terms in $\boldsymbol{i}$, $\boldsymbol{j}$ and $\boldsymbol{k}$, we then get equation (4.21a), equation (4.21b) and equation (4.21c), respectively, as required.

APPENDIX C

Solutions and hints for selected problems

Chapter 2

2.1. 5.3×10^{18} kg, 5.3×10^{21} J K^{-1}, 5.7×10^{24} J K^{-1}, 2.5 m.
2.2. 13.4 m.
2.3. 14.2 km, 19.5 km.
2.4. 30.7 km.
2.5. 9.76 K km^{-1}, 9.61 K km^{-1}.
2.6. 9.62 min.
2.8. 2.39×10^9 J m^{-2}. APE $= (2^{-1} - 2^{-(\kappa+1)}) c_p p_0 \Delta\theta / [g(\kappa + 1)]$.
2.10. (a) 300 hPa. (b) Dry: all. Saturated: above 780 hPa. (c) Hint: use the relationship between μ_s and T_d. 8.3 and 0.75 g kg. (d) 2 °C. (e) Hint: suppose that a DALR is established over time in the lowest levels of the atmosphere, moving upwards from the ground as shown in Figure C.1. At time t_1, parcels can rise adiabatically along the $\theta = \theta_1$ line, at a later time t_2 they can rise along $\theta = \theta_2$ and so on.
Suppose that at time t_3 the dry adiabatic has reached $\theta = \theta_3$, intersecting the environment curve at P, where the humidity mixing ratio = 8.3 g kg^{-1} curve also intersects the environment curve. Now, if condensation did not occur, a parcel rising adiabatically from the ground would retain its water-vapour mixing ratio (8.3 g kg^{-1}) as well as its potential temperature θ. At P the mixing ratio precisely

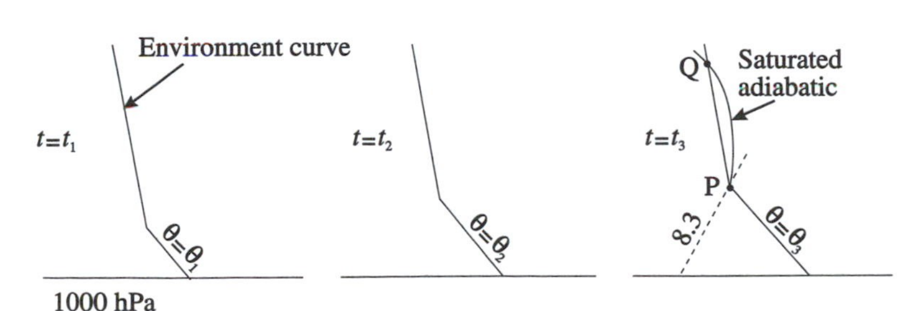

Figure C.1 Hints for Problem 2.10(e)

equals the saturation value at that point, so condensation occurs. P is the *lifting condensation level* and equals about 960 hPa in this question.

(f) From P the now-saturated parcel can rise along the saturated adiabatic, provided that it is positively buoyant, i.e., provided that its density is less than (and thus its temperature is greater than) that of the surroundings. It can therefore rise until the relevant saturated adiabatic crosses the environment curve again, at point Q (about 630 hPa in this question).

2.11. The mass of the layer ($\propto \Delta p$) is conserved, so the new pressure at the top is 600 hPa. The new temperature at the bottom is −8.3 °C and that at the top is −15.5 °C. The bottom of the second layer rises to 800 hPa and the new temperature is 0 °C.

2.12. The energy needed to raise a parcel to the first point is 7 J; the energy released between the first and second points is 55 J. The net energy released is 48 J.

2.13. 90.3 days, 21.7 h, 13 min. Re $\leq 10^{-5}$, 10^{-2} and 10, repectively, so Stokes's Law is not necessarily valid in the third case.

2.14. (i) 0.39 K, (ii) 23 min.

Chapter 3

3.1. (ii) 5800 K, 290 K. (iii) $B_\nu = 2.25\times10^{-8}\,\mathrm{W\,m^{-2}\,steradian^{-1}\,Hz^{-1}}$, $B_\lambda = 2.7\times10^{13}\,\mathrm{W\,m^{-2}\,steradian^{-1}\,m^{-1}}$, 8.6%.

3.2. (i) Both surfaces are black bodies; the temperature of ΔS_2 is sufficiently different from T_1 that it has negligible emission in the passband $\Delta\nu$; there is no absorption or scattering between the surface elements.
(ii) $F_s = 1381\,\mathrm{W\,m^{-2}}$. (iii) $\sigma T_1^4 \pi a^2$ (independently of h). (iv) The power absorbed by the sphere is $2\sigma T_1^4 \pi a^2$.

3.4. (i) 235 K, (ii) $\alpha F_s/\pi$, (iii) white sphere: 227 K, black sphere: 333 K.

3.5. (i) $\chi^* = 1.66k\mu p/[g(1+\mu)]$, (iv) $F_z = 183\,\mathrm{W\,m^{-2}}$, (v) 20.5 K.

3.7. At $\nu = \nu_0$, $\mathcal{T}_\nu = \exp\{-[\rho_a l S/(\pi\gamma_L)]\}$. This is independent of p if the mixing ratio $\rho_a/\rho = \mu =$ constant.

3.9. Pressure ≈ 4 hPa.

3.10. Last part: $\bar{p} = p(z)/2$.

3.11. (ii) $p_m = 1$ hPa, $h(p_m) = 21\,\mathrm{K\,day^{-1}}$, (iii) 280 K.

3.12. The reflection and transmission coefficients are given below.

χ_1^*	R	T	1 − R − T
10	0.4987	0.4983	0.002 994
20	0.6635	0.3305	0.005 970
30	0.7449	0.2462	0.008 920
∞	0.9255		

Chapter 4

4.1. The mean separation is $(n/V)^{-1/3} \sim 3\,\text{nm}$. Taking $\sigma \sim 0.3\,\text{nm}$, the mean free path $\sim 200\,\text{nm}$. The mean free path $\propto T/p$, where T is fairly constant, but p decreases exponentially with altitude; it is of the order of several metres at $p = 10^{-4}\,\text{hPa}$, i.e., over 100 km altitude.

4.4. 1.5 cm.

4.5. Approximate values of Ro: (a) 10, (b) 10^4, (c) 5×10^3.

4.6. For real solutions, we need $-dp/dr \le \rho\Omega^2 r$. Overestimate.

4.7. Assume that T_1 and $T_2 \simeq 280\,\text{K}$ and v_1 and $v_2 \simeq 10\,\text{m}\,\text{s}^{-1}$, say. Then $v_1T_2 - v_2T_1 \simeq (v_1 - v_2)T_2$. So $\alpha \simeq 0.6°$.

4.8. $71\,\text{m}\,\text{s}^{-1}$.

4.10. Warm sector, mean temperature $\simeq 272\,\text{K}$. $75\,\text{m}\,\text{s}^{-1}$.

Chapter 5

5.2. (a) 10 km, (b) 100 km.

5.3. Period $\simeq$ 120 min, time taken $\simeq$ 480 min, $\Delta X \simeq 2.3\,\text{km}$, $\Delta Z \simeq 115\,\text{m}$.

5.5. $\overline{P}/\overline{K} = (\omega^2 - f_0^2)/(\omega^2 + f_0^2)$.

5.6. 2.6 days.

5.7. (a) 13 000 km, (b) 23 000 km. Length of latitude circle at 60 °N = 20 000 km.

5.8. Cross-isobar flow is towards low pressure, in general. With $\pi h = 1\,\text{km}$, $\nu \sim 5\,\text{m}^2\,\text{s}^{-1}$. $w_\tau \sim 3\,\text{mm}\,\text{s}^{-1}$. The direction of stress is 45° anticlockwise from u_p, magnitude $\sim 0.07\,\text{N}\,\text{m}^{-2}$.

5.9. $2.7\,\text{m}\,\text{s}^{-1}$.

5.10. Hint: note that $\delta\rho = (\partial\rho/\partial y)\,\delta y + (\partial\rho/\partial z)\,\delta z$, where $\delta y = \delta s\cos\phi$, $\delta z = \delta s\sin\phi$. Angle $= 7.4\times10^{-4}$ radians.

5.11. 1.11 days, 7000 km, 4600 km.

Chapter 6

6.2. The change of Gibbs free energy is + 72.1 kJ mol^{-1}, so reaction cannot take place spontaneously.
6.4. $P_s = 6.1 \times 10^{21}$ photons m^{-2} s^{-1}. (i) 0.56, (ii) 4×10^{-3}, (iii) 1.7×10^{-3}.
6.7. O decays rapidly (on a time scale of a few seconds) after sunset, the amount of O_3 stays close to its sunset value. X decays, XO tends to a constant amount.

Chapter 7

7.3. (a) This is a water-vapour rotation band, indicating the presence of water vapour in the atmosphere. (b) This is the CO_2 vibration–rotation band. (c) North pole: surface temperature $\simeq$ 140 K (the freezing temperature of CO_2), atmospheric temperature approaches a constant 190 K higher up. South pole: surface temperature $\simeq$ 190 K, lower atmosphere $\simeq$ 230 K, upper atmosphere (spike in spectrum) 190–230 K. Midlatitudes: surface $\simeq$ 280 K, decreasing to 180 K in the upper atmosphere. (d) The terrestrial spectrum is more complicated, indicating the presence of more absorbers. The Earth has a warmer surface and atmosphere.
7.4. Transmission = 0.53. One can calculate the column ozone from this, if the extinction coefficient is both known and constant.
7.5. Transmittance = 0.9. Spectral radiance deficit = 4.8×10^{-14} W m^{-2} steradian^{-1} Hz^{-1}. Temperature error = −0.9 K.
7.6. (a) $\alpha = 1$, $\beta = k\mu/g$, (b) $\alpha = 2$, $\beta = S\mu\gamma_{L0}/[2g\pi(\nu - \nu_0)^2 p_0]$, (c) $\alpha = 1$, $\beta = 2[S\mu\gamma_{L0}/(\delta^2 g p_0)]^{1/2}$.
7.7. $p_m = \beta^{-1/\alpha}$, pressure width = $(2.68^{1/\alpha} - 0.23^{1/\alpha})p_m$. Case (b) is narrower.

Bibliography

Acheson, D. J. (1990). *Elementary Fluid Dynamics*. Oxford University Press.

Adkins, C. J. (1983). *Equilibrium Thermodynamics*. Cambridge University Press, third edition.

Andrews, D. G. (1991). A stratospheric transport system. *Phys. World*, **4**(11), 41–6.

Andrews, D. G., Holton, J. R. and Leovy, C. B. (1987). *Middle Atmosphere Dynamics*. Academic Press.

Atkins, P. W. (1994). *Physical Chemistry*. Oxford University Press, fifth edition.

Balsley, B. B. and Gage, K. S. (1980). The MST radar technique: potential for middle atmospheric studies. *Pure Appl. Geophys.*, **118**, 452–93.

Balsley, B. B., Ecklund, W. L. and Fritts, D. C. (1983). VHF echoes from the high-latitude mesosphere and lower thermosphere: observations and interpretations. *J. Atmos. Sci.*, **40**, 2451–66.

Banwell, C. N. and McCash, E. M. (1994). *Fundamentals of Molecular Spectroscopy*. McGraw-Hill, fourth edition.

Boas, M. L. (1983). *Mathematical Methods in the Physical Sciences*. Wiley, second edition.

Bohren, C. F. (1987). *Clouds in a Glass of Beer*. Wiley.

Bohren, C. F. and Albrecht, B. A. (1998). *Atmospheric Thermodynamics*. Oxford University Press.

Bolton, D. (1980). The computation of equivalent potential temperature. *Mon. Wea. Rev.*, **108**, 1046–53.

Born, M. and Wolf, E. (1980). *Principles of Optics*. Pergamon, sixth edition.

Bransden, B. H. and Joachain, C. J. (1989). *Introduction to Quantum Mechanics*. Longman.

Brasseur, G. and Solomon, S. (1986). *Aeronomy of the Middle Atmosphere*. Reidel, second edition.

Brewer, A. W. (1949). Evidence for a world circulation provided by the measurements of helium and water vapour distribution in the stratosphere. *Quart. J. Roy. Meteorol. Soc.*, **75**, 351–63.

Chamberlain, J. W. and Hunten, D. M. (1987). *Theory of Planetary Atmospheres*. Academic Press, second edition.

Chapman, S. (1930). A theory of upper atmospheric ozone. *Mem. Roy. Meteorol. Soc.*, **3**, 103–25.

Charney, J. G. and Drazin, P. G. (1961). Propagation of planetary-scale disturbances from the lower into the upper atmosphere. *J. Geophys. Res.*, **66**, 83–109.

DeMore, W. B., Sander, S. P., Golden, D. M., Hampson, R. F., Kurylo, M. J., Howard, C. J., Ravishankara, A. R., Kolb, C. E. and Molina, M. J. (1997). Chemical kinetics and photochemical data for use in stratospheric modeling, Evaluation

Number 12. JPL Publication 97–4, Jet Propulsion Laboratory.

Dobson, G. M. B. (1956). Origin and distribution of the polyatomic molecules in the atmosphere. *Proc. Roy. Soc.* A, **236**, 187–93.

Dobson, G. M. B. and Normand, C. W. B. (1957). Observers' handbook for the ozone spectrophotometer. *Ann. Int. Geophys. Year*, **5**, 46–81.

Faber, T. E. (1995). *Fluid Dynamics for Physicists*. Cambridge University Press.

Finn, C. B. P. (1993). *Thermal Physics*. Chapman and Hall, second edition.

Fleming, E. L., Chandra, S., Barnett, J. J., and Corney, M. (1990). Zonal mean temperature, pressure, zonal wind and geopotential height as functions of latitude. *Adv. Space Res.*, **10**(12), 11–59. For data tables of the COSPAR Middle Atmosphere Reference Atmosphere, see pages 357–517.

Gardiner, B. G. (1989). The Antarctic ozone hole. *Weather*, **44**, 291–8.

Gill, A. E. (1982). *Atmosphere–Ocean Dynamics*. Academic Press.

Goody, R. M. (1995). *Principles of Atmospheric Physics and Chemistry*. Oxford University Press.

Goody, R. M. and Yung, Y. L. (1989). *Atmospheric Radiation*. Oxford University Press, second edition.

Graedel, T. E. and Crutzen, P. J. (1995). *Atmosphere, Climate, and Change*. Scientific American Library.

Grant, I. S. and Phillips, W. R. (1990). *Electromagnetism*. Wiley, second edition.

Guenault, A. M. (1995). *Statistical Physics*. Chapman and Hall, second edition.

Haltiner, G. J. and Williams, R. T. (1980). *Numerical Prediction and Dynamic Meteorology*. Wiley, second edition.

Hanel, R. A., Schlachman, B., Rogers, D. and Vanous, D. (1971). Nimbus 4 Michelson interferometer. *Appl. Optics*, **10**, 1376–82.

Hanel, R. A., Conrath, B. J., Jennings, D. E. and Samuelson, R. E. (1992). *Exploration of the Solar System by Infrared Remote Sensing*. Cambridge University Press.

Hartmann, D. L. (1994). *Global Physical Climatology*. Academic Press.

Herzberg, G. (1945). *Molecular Spectra and Molecular Structure. II. Infra-red and Raman Spectra of Polyatomic Molecules*. Van Nostrand.

Herzberg, L. (1965). Solar optical radiation and its role in upper atmospheric processes. In C. O. Hines, I. Paglis, T. R. Hartz and J. A. Fejer, editors, *Physics of the Earth's Upper Atmosphere*, pages 31–45. Prentice Hall.

Hide, R. and Mason, P. J. (1975). Sloping convection in a rotating fluid. *Adv. Phys.*, **24**, 47–100.

Hobbs, P. V. (1995). *Basic Physical Chemistry for the Atmospheric Sciences*. Cambridge University Press.

Holton, J. R. (1992). *An Introduction to Dynamic Meteorology*. Academic Press, third edition.

Holton, J. R., Haynes, P. H., McIntyre, M. E., Douglass, A. R., Rood, R. B. and Pfister, L. (1995). Stratosphere–troposphere exchange. *Rev. Geophys.*, **33**, 403–39.

Houghton, J. T. (1986). *The Physics of Atmospheres*. Cambridge University Press, second edition.

Houghton, J. T., Taylor, F. W. and Rodgers, C. D. (1984). *Remote Sounding of Atmospheres*. Cambridge University Press.

Houghton, J. T., Meira Filho, L. G., Callender, B. A., Harris, N., Kattenberg, A. and Maskell, K., editors (1996). *Climate Change 1995. The Science of Climate Change*. Cambridge University Press. (Contribution of Working Group I to the Second Assessment Report of the Intergovernmental Panel on Climate Change.)

Jackson, J. D. (1999). *Classical Electrodynamics*. Wiley, third edition.

Kaye, G. W. C. and Laby, T. H. (1986). *Tables of Physical and Chemical Constants*. Longman, fifteenth edition.

Landau, L. D. and Lifshitz, E. M. (1980).

Statistical Physics, Part 1. Pergamon, third edition.

Lary, D. J. (1997). Catalytic destruction of stratospheric ozone. *J. Geophys. Res.*, **102**, 21 515–21 526.

Lean, J. (1987). Solar ultraviolet irradiance variations: a review. *J. Geophys. Res.*, **92**, 839–68.

Li, D. and Shine, K. P. (1995). A 4-dimensional ozone climatology for UGAMP models. U.K. Universities' Global Atmospheric Modelling Programme internal report No. 35.

Lide, D. R., editor (1995). *Handbook of Chemistry and Physics*. CRC Press, seventy-sixth edition.

Liou, K. (1980). *An Introduction to Atmospheric Radiation*. Academic Press.

London, J. (1980). Radiative energy sources and sinks in the stratosphere and mesosphere. In A. Aiken, editor, *Proceedings of the NATO Advanced Study Institute on Atmospheric Ozone: its Variation and Human Influences*, pages 703–21. Report FAA–EE–80–20.

Lorenz, E. N. (1955). Available potential energy and the maintenance of the general circulation. *Tellus*, **7**, 157–67.

Lorenz, E. N. (1963). Deterministic nonperiodic flow. *J. Atmos. Sci.*, **20**, 130–41.

Lorrain, P., Corson, D. P. and Lorrain, F. (1988). *Electromagnetic Fields and Waves*. Freeman, third edition.

Lynch, D. K. and Livingston, W. (1995). *Color and Light in Nature*. Cambridge University Press.

Lyons, L. (1995). *All You Wanted to Know About Mathematics But Were Afraid to Ask*, volume 1. Cambridge University Press.

Mandl, F. (1988). *Statistical Physics*. Wiley, second edition.

McIlveen, R. (1991). *Fundamentals of Weather and Climate*. Chapman and Hall.

McIntosh, D. H. and Thom, A. S. (1983). *Essentials of Meteorology*. Taylor and Francis.

McMurry, S. M. (1993). *Quantum Mechanics*. Addison-Wesley.

Meier, R. R. (1991). Ultraviolet spectroscopy and remote sensing of the upper atmosphere. *Space Sci. Rev.*, **58**, 1–185.

Pedlosky, J. (1987). *Geophysical Fluid Dynamics*. Springer-Verlag, second edition.

Rae, A. I. M. (1992). *Quantum Mechanics*. Institute of Physics, third edition.

Read, P. L. (1993). Applications of chaos to meteorology and climate. In T. Mullin, editor, *The Nature of Chaos*, pages 222–60. Oxford University Press.

Rodgers, C. D. and Walshaw, C. D. (1966). The computation of infra-red cooling rate in planetary atmospheres. *Quart. J. Roy. Meteorol. Soc.*, **92**, 67–92.

Rogers, R. R. and Yau, M. K. (1989). *A Short Course in Cloud Physics*. Pergamon, third edition.

Salby, M. L. (1996). *Fundamentals of Atmospheric Physics*. Academic Press.

Solomon, S. (1990). Progress towards a quantitative understanding of Antarctic ozone depletion. *Nature*, **347**, 347–54.

Stephenson, G. (1973). *Mathematical Methods for Science Students*. Longman, second edition.

Stratospheric Ozone (1996). Report of the United Kingdom Stratospheric Ozone Review Group. Prepared at the request of the Department of the Environment.

Tabor, D. (1991). *Gases, Liquids and Solids*. Cambridge University Press, third edition.

Thomas, L. (1987). Laser radar observations of middle-atmosphere structure and composition. *Philos. Trans. Roy. Soc.* A, **323**, 597–609.

Thorne, A. P. (1988). *Spectrophysics*. Chapman and Hall, second edition.

Thuburn, J. and Craig, G. C. (1997). GCM tests of theories for the height of the tropopause. *J. Atmos. Sci.*, **54**, 869–82.

Trenberth, K. E., editor (1992). *Climate System Modeling*. Cambridge University Press.

Tritton, D. J. (1988). *Physical Fluid Dynamics*. Oxford University Press, second edition.

U. S. Standard Atmosphere (1976). National Oceanic and Atmospheric Administration, National Aeronautics and Space Administration, U. S. Air Force. U. S. Government Printing Office.

Vaughan, G., Wareing, D. P., Jones, S. B., Thomas, L. and Larsen, N. (1994). Lidar measurements of Mt. Pinatubo aerosols at Aberystwyth from August 1991 through March 1992. *Geophys. Res. Lett.*, **21**, 1315–18.

Wallace, J. M. and Hobbs, P. V. (1977). *Atmospheric Science*. Academic Press.

Wayne, R. P. (2000). *Chemistry of Atmospheres*. Oxford University Press, third edition.

Index